The ongoing process of revision and rethinking of the foundations of economic theory leads to great complexities and contradictions at the heart of economics. *The Economics of Innovation* provides a fertile challenge to standard economics and one that can help it overcome its many criticisms.

This authoritative book from Cristiano Antonelli provides a systematic account of recent advances in the economics of innovation. By integrating this account with the economics of technological change, the book elaborates an understanding of the paths and the sequence of determinants and effects of the introduction of new technologies.

Many within the innovation economics community will appreciate this excellent, comprehensive account, provided by a respected expert, but it is a book that also needs to be read by *all* those with an interest in economic theory.

Cristiano Antonelli is Professor of Economics at the University of Turin, Italy. He is the managing editor of the *Economics of Innovation and New Technology*, and has written numerous books including *The Microdynamics of Technological Change* (Routledge, 1999).

Studies in global competition
Edited by John Cantwell, University of Reading, UK and David Mowery,
University of California, Berkeley, USA

The Economics of Innovation, New Technologies and Structural Change

Cristiano Antonelli

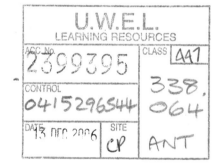

Routledge
Taylor & Francis Group

LONDON AND NEW YORK

First published 2003
by Routledge
2 Park Square, Milton Park, Abingdon, Oxon, OX14 4RN

Simultaneously published in the USA and Canada
by Routledge
270 Madison Ave, New York NY 10016

Routledge is an imprint of the Taylor & Francis Group

Transferred to Digital Printing 2006

© 2003 Cristiano Antonelli

Typeset in Times by
Keystroke, Jacaranda Lodge, Wolverhampton

British Library Cataloguing in Publication Data
A catalogue record for this book is available from the British Library

Library of Congress Cataloging in Publication Data
Antonelli, Cristiano.
   The economics of innovation, new technologies and structural change /
Cristiano Antonelli.
   p. cm. – (Studies in global competition ; 14)
Includes bibliographical references and index.
1. Technological innovations–Economic aspects. 2. Competition,
International. I. Title. II. Series.

HC79.T4 A582 2002
338′.064–dc21   2002031737

ISBN10: 0–415–29654–4 (hbk)
ISBN10: 0–415–40643–9 (pbk)

ISBN13: 978–0–415–29654–0 (hbk)
ISBN13: 978–0–415–40643–7 (pbk)

Contents

Figures

Foreword

This book provides an enlarged Schumpeterian analysis of the implications of the introduction of a new technological system such as new information and communication technology in the global economy. New information and communication technologies are the result of a complex innovation process and as such can be considered a general purpose technology which has a wide spectrum of applications with important effects in terms of potential growth of total factor productivity and significant bias in the use of production factors across countries and regions in the global economy.

Their introduction and use challenge both economics and the economy at large, especially when the global economy is considered. In order to understand their effects and ultimate consequences, in terms of divergent rates of growth across industries, countries and regions within countries, an effort is necessary to bring together different traditions of analysis from Schumpeterian economics to standard analysis of the economics of technical change.

The analysis of the long-term process of generation and introduction of new information and communication technologies and of their economic effects at large calls for a comprehensive assessment of the different approaches elaborated in economics to cope with the analysis of the determinants and consequences of the introduction of technological innovations. This analysis suggests that one must pay special attention to the differentiated endowments of regions and countries and to the variance of relative prices of production factors. The matching of such analysis together with the understanding of the innovation process itself delivers important results in terms of assessment of the asymmetric effects of the introduction and adoption of new technologies and hence the basic recipes for public policy interventions.

The book has a strong European flavor and provides a tentative interpretation of the delays and problems the European economy must face in order to take full advantage of new information and communication technologies. So it can become a reference book to help understand the problems of coping with a new radical general purpose technology, mainly elaborated elsewhere (i.e. the US).

The dynamic approach of the analysis also makes clear that the introduction of new technologies and their diffusion is a process which takes place in out-of-equilibrium conditions to which agents are able to react with the introduction of innovations fed with self-generating effects.

This book is the result of an ongoing process of revision and rethinking of the foundations of economic theory and specifically of microeconomics. The foundations of economics seem at the same time more fragile and yet more fertile than it is too often acknowledged. Heterodox economists on the one hand stress the fragility of many assumptions and suggest that much economic analysis should be abandoned. On the other hand, orthodox economics seems often too little aware of the fragility of many assumptions. Too much reliance is placed upon a set of basic assumptions that have only a limited scope of application.

The methodological choice on which this book elaborates consists exactly in the opposite combination. Microeconomics is fragile and many basic tools hold only in a very specific set of circumstances. Yet it is a fertile field of analysis. Small changes of some subsets of assumptions in fact can yield important results, especially if the rigorous and deductive chains of arguments which characterizes economic analysis are used and implemented.

The joint didactics, carried on for many years, of an orthodox course of microeconomics together with a lively class of economics of innovation, where much material is drawn from heterodox approaches, is surely responsible for these outcomes. The teaching of economics of innovation provides much stimulation and insight in reconsidering the basic assumptions of economic analysis. The static limitations of the foundations of economics are put under strain by the dynamic approach of economics of innovation. The teaching of the foundations of economics induces the temptation to refill old bottles. The process becomes a fertile challenge. The systemic strength of standard economics makes it possible to push further, in a rigorous way, many insights and suggestions provided by economics of innovation.

The basic aim is to contribute the analysis of the economy as an open system, while building upon the tools elaborated to study the economy as a closed system. Once more, an effort to stand on giants' shoulders.

Acknowledgments

Preliminary versions of different chapters and sections have been presented in different contexts. The first version of the basic argument was presented at the conference in Honor of Paul David "New Frontiers in the Economics of Innovation and New Technology" organized by the Laboratorio di economia dell'innovazione "Franco Momigliano" and the Dipartimento di Economia of the Università di Torino at the Accademia delle Scienze in Torino (May 2000). Further versions and sections have been subsequently presented at the European Summer School on Industrial Dynamics, in Cargese (September 2000); the conference "Innovation, Time and Space" jointly organized by the Center for Technology, Innovation and Culture (TIK) of the University of Oslo and the Department of Economics of the University of Urbino, in Urbino (October 2000); the workshop on Cognitive Economics organized by the Center for Cognitive Economics of the University Amedeo Avogadro, in Torino and Alessandria (November 2000); the Dialogue Workshop of the European Commission "Improving the Socio-economic Knowledge Base. The Regional Level of Implementation of Innovation and Education and Training Policies", in Brussels (November 2000); the conference "Tecnologia e società" organized by the Accademia dei Lincei in Rome (December 2000); the ESRC International Research workshop "Innovation and Competitive Cities in the General Economy", organized by Oxford Brookes University in Oxford at Worcester College (March 2001); the colloquium "Nouvelle Economie: Théories et Evidences" organized by the Université Jean Monnet, in Sceaux (May 2001); the workshop "New Policy Rationales for the Public Funding of Science" organized by the SPRU of the University of Sussex in Paris at the Ecoles des Mines (May 2001); the conference of the Comité d'orientation de l'Observatoire EPFL Science, Politique et Société of the Ecole Fédérale Polytechnique de Lausanne (September 2001); the kick-off meeting of the research projects and thematic networks selected under the Key Action "Improving the socio-economic knowledge base" with the project "Technological Knowledge and Localised Learning: What Perspectives for a European Policy" in Brussels (September 2001); the XXVth Convegno nazionale di Economia e politica Industriale "Venticinque anni di industria. Organizzazione industriale, corporate governance e crescita di impresa italiana" in Bologna (September 2001); the workshop on the "Economics of Knowledge" organized in Lyon by the Centre Auguste et Leon Walras at the Institut des Sciences de l'Homme (CNRS-Université Lyon 2) (October 2001); the Conference of the Accademia delle Scienze at the Politecnico di Milano "Sviluppo delle tecnologie dell'informazione e della comunicazione (ict) e 'new economy'" (November 2001); the Raffaele

Mattioli Lecture "Globalization, Productivity Change and the Economics of Information" in Milano at the Università Bocconi (November 2001); the workshop "Reappraising Production Theory: Concepts, Cases and Models" organized by the Max Planck Institute for Research into Economic Systems in Jena (November 2001); the Conference "Innovation and Growth: A New Challenge for Regions" organized by the IDEFI of the Université de Nice-Sophia Antipolis and the Département des Etudes of OFCE in Sophia Antipolis (January 2002); the conference "Percorsi teorici e progetti didattici nell'economia dell'innovazione e della tecnologia" organized at the Fondazione Giovanni Agnelli in Torino (June 2002); the XIVth Biennal Conference of the International Telecommunications Society "Challenges and Opportunities in the Digital Century: The Role of Information and Telecommunications" in Seoul (August 2002); "Localized Technological Change and Path Dependence: The Role of Relative Factors Prices" at the CRIC of the University of Manchester (September 2002).

The role of ETE, the European Innovation Network "Economic Transformation of Europe" promoted by the Center for Research in Innovation and Competitiveness of the Victoria University of Manchester, has been especially important, with meetings organized at the CRIC of the University of Manchester (March 2001), at the Chalmers University of Technology in Göteborg (July 2001) and at the Max Planck Institute for Research into Economic Systems in Jena (February 2002).

The useful comments of Mario Amendola, Mario Calderini, Maurice Catin, Uwe Cantner, Robin Cowan, Paul David, Aldo Enrietti, Martin Fransman, Jean Luc Gaffard, Pierre Garrouste, Bernard Gulhon, Christian Le Bas, Giorgio Lunghini, Roberto Marchionatti, Stan Metcalfe, Keith Pavitt, Mario Pianta, Alberto Quadrio Curzio, Michel Quéré, Marco Vivarelli, Simon Teitel, Nick von Tunzelman, Antonello Zanfei and Ulrich Witt, among others, are gratefully acknowledged. The remarks of Aldo Geuna, Pier Paolo Patrucco, Francesco Quatraro, Andrea Mina and Martin Marchesi have been especially appreciated. Finally, I acknowledge the detailed comments of the three anonymous referees. The usual disclaims apply.

The research was conducted within the research context and with the support of the competitive research grants awarded for the years 2001 and 2002 by the National Research Funds of the Italian Ministry for Education and Research (MIUR) to the National project "The impact of technological innovation and globalization on the economic performances of Italy and Europe" of the National Research Pool of the Universities of Torino, Bologna, Camerino and Urbino, coordinated by Professor Antonello Zanfei of the University of Urbino.

The financial support of the Research Funds of the University of Torino (Department of Economics: research grants for the years 2000, 2001 and 2002) of the Fondazione Giovanni Agnelli and of the Compagnia di San Paolo is gratefully acknowledged.

A large portion of the work was conducted with the help of funding from the European Union Directorate for Research, to the Fondazione Rosselli, in the context of the Key Action "Improving the socio-economic knowledge base" as a part of the project "TELL" (Technological Knowledge and Localised Learning: What Perspectives for a European Policy?) carried on under the research contract No. HPSE-CT2001-00051.

1 Introduction

Economics of innovation has grown rapidly in the last part of the twentieth century around a problematic core: the puzzle of total factor productivity growth. The continual growth of output and efficiency experienced in most countries since industrialization cannot be explained only in terms of an increase of inputs used in the production process. The introduction of innovations and new technologies plays a major role in changing the efficiency of use of inputs and hence output levels per unit of input.

After much progress, economics of innovation seems more and more to concentrate on the effort to elaborate a theory of economic creativity with a strong impact on the theory of the firm and the theory of organization and decision making.

In parallel, mainstream economics has been able to appropriate many results of the economics of innovation with the successful development of the so-called "new growth theory". New growth theory has been able to assimilate many achievements of the economics of innovation such as the distinctions between tacit and codified knowledge, the analysis of appropriability regimes, the understanding of the role of technological externalities, spillovers and systems of innovations. New growth theory however rejects all the complementary implications of the economics of innovation in terms of the disequilibrium conditions that are the necessary conditions to understand the dynamics of introduction of technological change and its effects. An equilibrium theory of growth, based upon the joint production of appropriable knowledge and relevant technological externalities, now provides an important achievement of mainstream economics. Firms innovate relying on the appropriability of specific knowledge and in so doing generate generic knowledge which spills over and is freely available. The introduction of innovations takes place smoothly in a monopolistic competition context with an increasing variety of products (Aghion and Howitt 1998).

Such a result seems to be based upon a highly selective and indeed highly partial recombination of equilibrium analysis with some key findings of the economics of innovation. The growing exclusive concern of economics of innovation with the nanoeconomics of creativity and knowledge seems to leave more and more room for such attempts.

An effort to build a more general framework of analysis in which the real achievements of economics of innovation can be valued seems necessary.

Innovation and disequilibrium cannot be decoupled, nor can the introduction of novelty and the continual and endogenous interplay between technological and structural change be reduced to a smooth process of introduction of an ever increasing variety of products.

A better understanding of the determinants and effects of technological change is possible only when the full set of structural elements of the economic system into which each new technology is being introduced is properly taken into account. The full array of direct and indirect dynamic interactions which characterize the introduction of a new technology into an economic system can be properly appreciated only when the structure of underlying relations among each component of the system is considered.

An approach is needed by which where the variety of innovations and the heterogeneity of markets and actors, both on the supply and the demand side, is recognized and yet their interdependence is emphasized. To do so the merging of the Schumpeterian and the Marshallian traditions of analysis is most useful. The latter focuses the heuristic power of partial equilibrium analysis but in a systemic context. The former calls attention to the role of the dynamic forces, brought about by innovation, in the perpetual unfolding of new perspectives and evolving forms of coherence among each element and subset of elements which are part of the system.

Although economic change keeps the system away from a stable equilibrium, a systemic analysis is necessary to understand the full set of laws of motion, including feedbacks and reactions, which characterize the sequence of events which lead to and parallel the generation, introduction, adoption and diffusion of new technologies. From this viewpoint the important progress made by the economics of innovation needs to be better integrated into the broader and more complex framework of analysis provided by the economics of technical change.

Such an effort seems all the more relevant and useful in the context of the new global economy, as it emerged and consolidated in the last decades of the twentieth century. The distinctive feature of the global economy in fact is the great variety of factors markets and yet single products markets where competitors based in different countries face each other with different technologies and different absolute and relative factors costs. In this context a basic tenet of economic analysis finds a new and problematic application. Not only do relative prices guide the choice of techniques, but also, and surprisingly, the choice of technologies.

Economics of innovation must cope with such a context and possibly find, in so doing, a much broader scope of analysis. The traditional borders between technical choice and technological change become in fact more and more blurred so that the divide between the economics of technical change and the economics of new technologies cannot longer hold.

A better integration of the economics of innovation with the economics of technical progress can yield important results so as to pave the way to a broader scope of analysis, one which is able to provide a systemic analysis of technological change where the microeconomic and the aggregate analysis integrate.

At any point in time the levels of output and total factor productivity are sensitive to the relative scale of the production factors. All changes in the composition of production factors has a direct effect on output and total factor productivity. The composition of inputs changes either when technology or the relative price of inputs change. Input composition affects output levels and average costs: hence standard procedures to assess total factor productivity growth both synchronically and diachronically.

Two new dimensions of technological change emerge, respectively general new technologies and contingent technological change. A general new technology can be identified when it consists of a neutral shift effect, such that all the map of isoquants slides towards the origin. The shift effect, moreover, is so consistent that even perspective users active in different factors markets will find the new technology superior to the previous one. The total factor productivity increase in this case is absolute. A contingent technological change is defined by the sheer change in the direction of new technologies, that is in the shape of the isoquants, without any actual increase in the absolute efficiency. A contingent technological change has a limited scope of application because it makes it possible total factor productivity increases only for specific combinations of factors costs.

This distinction between the notions of general and contingent technological change makes it possible to better appreciate our understanding of innovation at the firm level as well as the assessment of the interplay between technological and structural change at the system level. The levels and the dynamics of relative prices of both basic and intermediary inputs, as determined by the original endowments and the industrial structure of the system and their changes, are the primary factors in the inducement mechanism which provides both incentives for and constraints on the rates and the direction of technological change. The rate and the direction of technological change are induced by the specific characteristics of the industrial and economic structure of the system at each point in time and by their changes. Specifically the levels of the relative prices induce the direction of the new technologies while their changes induce the rates of introduction of innovations.

Schumpeterian growth cycles associated with the introduction of new biased technologies can be explained in terms of the market dynamics in upstream markets for production factors. Regional integration of economic systems and internal competition are affected by the interplay of relative prices and technological change. The intentional reshaping of relative prices can become the basic guideline of a growth-oriented industrial policy to take advantage of the composition effects.

In turn, however, the rate and the direction of technological change introduced and adopted at each point in time has a direct bearing on the structure of relative prices, with a strong and recursive path-dependent process of localized causation where technological change and structural change affect each other.

The understanding of the interactions between the structure of the economic system, especially in terms of relative prices and hence the characteristics of the markets for intermediary and basic inputs, and the characteristics of a new technology in terms of both the advance of general efficiency and the changes in the intensity of usage of production factors, provides a fertile and rich context into which the analysis of the determinants and effects of the generation, introduction, adoption and diffusion of new technologies can be developed.

In this context technological variety can be found with significant effects both at the macroeconomic and system levels and with respect to our actual understanding at the microeconomic levels of the behavior of firms. It is interesting to note that the notion of technological substitution finds some preliminary elements in the debate about reswitching. The debate about reswitching has been very sharp and absorbed much attention. Its focus, however, was more about the implications of the overlapping of two set of techniques with respect to the theoretical definition of the equilibrium levels of profits and wages in a general equilibrium analysis and growth theory (Pasinetti 1962). The acquisitions of the debate on reswitching found little application in the economics of technological change. As a matter of fact, however, the debate on reswitching anticipated the analysis of techniques and technologies which cannot be defined in absolute terms as either progressive or regressive (Heertje 1973).

The actual ranking of new technologies, in terms of both price and output efficiency, becomes much less clear when a variety of factors markets is considered. For each technology, traditionally defined as an envelope of equivalent techniques, output and average costs are sensitive to the relative prices of production factors and to their scale of usage.

As Vernon Ruttan remarks: "if technical change is nonneutral between t_0 and t_1 the measure of technical change is no longer independent of the relative prices that prevail" (Ruttan 2001: 56).

In this context the hypothesis of a variety of agents and behaviors in economic analysis has many major implications. At a time of increasing globalization, the hypotheses of a substantial heterogeneity of local factors markets in each country seem more and more compelling and realistic. Such hypotheses are all the more plausible at a time of rapid and biased technological change, characterized by strong factor-saving and factor-using effects brought by the introduction of a new general purpose technology such as the recent waves of innovation in new information and communication technologies. Because of the sensitivity of output and average costs to the relative prices of production factors, a variety of "efficient" technologies can be found across global markets. In this context the issues of technological variety and contingent technological advance become relevant.

When factors markets heterogeneity in terms of prices is admitted, comparative analysis of the performances of a given technology in different regions becomes a very difficult task. By the same token the assessment of the actual performance of a new technology changes along with any change in the relative cost of production factors. The actual effects the technology are contingent upon the systems of relative factor prices in each local market. The levels of total factor productivity associated with a single technology may vary across regions, according to the relative prices of production factors. This adds to the consequences of the sensitivity of production costs, and hence in a global competitive market, of output levels to the ratio of factors costs. Such effects are all the more relevant as they reinforce each other. They apply synchronically to differences in the relative costs of inputs across industries and regions, in assessing the actual levels of increase of total factor productivity when a single new technology is introduced in different regions and diachronically when changes take place along historic time, in factors markets in assessing, with a given technology, the levels of production costs.

Technological changes can no longer be ordered just in terms of total factor productivity levels: relative prices of production factors matter in assessing the actual bottom line average costs, according to the relative productivity and price of each factor. Standard procedures to assess total factor productivity growth can appreciate to some extent the effects of all synchronic and diachronic changes in factor costs.

These issues seem extremely relevant at the microeconomic level of analysis and have important effects on our actual understanding of both the behavior of firms and the aggregate dynamics of the economic system. The differences in the production functions at work and in the factors markets, among industries and regions, are much sharper than at the aggregate levels. The actual assessment of the real evolution of technological change at a microeconomic, industrial and regional level is much more complex than at the aggregate levels and important dynamic implications need to be taken into account.

The understanding of the reciprocal feedbacks between the structure of each economic system, integrated in the global economy, and the rate and the direction of technological change makes it possible to grasp the basic path-dependent character of growth and change. The understanding of the path-dependent interaction between structural change and technological change seems a major contribution made possible by the integration of the economics of innovation tradition of analysis and the economics of technical change.

The rest of the book is structured in three parts as follows. Part I recalls the building blocks on which the analysis of the recursive interactions between innovation, new technologies and structural change is conducted. In this part Chapter 2 provides a general review of the main achievements of economics of innovation. The key role of out-of-equilibrium analysis in solving the puzzle of total factor productivity growth is elaborated and the complementarity between disequilibrium and the very notion of innovation is articulated.

Chapter 3 retrieves the important achievements of the economics of technical change as it emerged from the 1930 through the 1960s and lays down the background analysis of the opportunities for merging the traditions of analysis of the economics of innovation and of the economics of technical change respectively. The strength of this latter tradition of analysis, with respect to the role of prices and especially relative prices in assessing productivity levels and the circular relation between production and distribution theory, is stressed.

Part II starts with Chapter 4 which elaborates the dynamic implications of the sensitivity of total factor productivity and average costs levels to equilibrium input composition in an economics of innovation perspective of analysis. Chapter 5 shows the important opportunities created by the merging of the results of economics of innovation with the tradition of the economics of technical progress in a disequilibrium context. The distinction between the inducement of innovation activated by all changes in the levels of absolute factors prices, of relative factors prices and of demand, and the inducement of the direction of technological change is introduced and elaborated. Chapter 6 explores the interaction between industrial dynamics and the introduction of new general and contingent technologies. The vertical effects in the relations among industries are analyzed as well as the horizontal effects within industries. The role of industrial dynamics in the evolution of the relative prices of intermediary inputs is stressed as a key element in technological choice, consisting of both the induced introduction of new technologies and the adoption of existing ones. Chapter 7 highlights the interactions between the markets for basic inputs with special attention to labor markets and the technological choices of firms. The discontinuity brought about by the composition effects are analyzed as well as the consequences of the regional integration of economic systems characterized by significant differences in factors endowments. Chapter 8 generalizes the analysis of the interdependence between the dynamics of economics structures and technological change in the global economy. Chapter 9 provides a simple analysis of the effects of the heterogeneity of consumers on the oligopolistic rivalry among producers that are able to introduce product innovations and to retain a direct control of a fraction of the general demand for a class of goods that are only partly substitutes. Niche prices can help firms to survive in single products markets and make profits even if they are less cost-effective than competitors. Such cost asymmetries in fact depend upon the adverse relative endowments of most productive production factors which induce a mismatch between the local factors markets and the bias of the new production functions, brought about by highly productive general purpose technologies. Chapter 10 emphasizes the recursive and path-dependent relationship between structural change and technological change in terms of the incentives and constraints it engenders for the generation, introduction, adoption and diffusion of innovations at the firm and the system level.

The analytical tools introduced in Chapters 4–10 are applied in Part III where the policy implications are also developed. Chapter 11 applies the framework so far elaborated to the understanding of the economics of new information and communication technology in Chapter 10. The policy implications are analyzed in Chapter 12. The understanding of the key role of relative prices provides a new scope for a growth-oriented economic policy. The coevolution between the structure of relative prices and the direction of technological change makes it possible to stress the notion of potential productivity growth, defined as the scope for measured productivity growth which can be obtained by appropriate changes in the system of relative prices. The conclusions summarize the main results of the analysis.

Part I
The building blocks

2 Shifting heuristics in the economics of innovation

Introduction

Economics of innovation is a recent fruitful area of specialization in microeconomic theory. During the last forty years, economics of innovation has emerged as a distinct area of enquiry at the crossing of industrial organization, regional economics and the theory of the firm, and as a well-identified area of specialization in the microeconomics of growth. In this context the interaction with sociology, philosophy, management science, biology and especially history has been a constant source of inspiration, and has provided new heuristic metaphors to apply in the context of the microeconomic analysis of new knowledge and new technologies and their effects with a strong emphasis on the study of the dynamics of growth and change.

After the discovery of the residual, such a process has been nurtured by the sequential and yet overlapping articulation of three wide-ranging heuristic metaphors: the manna, the trajectory and the network. Each of them has made it possible to achieve important analytical results, which have enabled stylized facts, analytical prospects and sometimes theoretical revelations to be elaborated. Today these approaches are competing and yet cooperating so as to make this discipline a particularly fertile and creative area of economic theory.

From this viewpoint, economics of innovation and new technologies is more than a new attempt by economics to extend its methodology into the realm of social sciences, such as it is in the case of the economics of education, health or risk, just to name a few new areas of specialization of economics. Economics of innovation and new technologies is also and mainly a necessary switchboard of interface and interaction with other social sciences, and is able to analyze the role of unexpected events in economic life.

The rest of the chapter provides a survey of the main steps through which economics of innovation has emerged. Pages 4–5 identify the birth of this area of specialization with the discovery of the residual. Pages 6–9 show the progressive reduction of the heuristic power of the notion of technological change as an exogenous process encapsulated in the metaphor of the manna. The manna leaves the pace to the notion of technological opportunity. Pages 9–18 review the biological grafts onto the merging economics of innovation provided by the application of the literature on epidemics to understand the diffusion of innovation and the life cycle approach to frame the sequence of events which follow the introduction of a new product. Section 5 presents the emergence of the metaphors of technological trajectory and technological paradigms, eventually substituted with the more articulated understanding of the role of path dependence and cumulativeness in technological change. Pages 18–24 introduce the notions of collective knowledge, networks of innovators and relevant attributes of technological change such as complementarity and interdependence as developments of the notion of non-appropriability and externality. Pages 24–25 presents the new challenges of the economics of innovation raised by the bifurcation between the foray of the new growth theory and the new microeconomics of knowledge. The new theory of growth provides an equilibrium explanation for the process of generation of new technological knowledge and introduction of technological innovation based upon the distinction between generic inappropriable knowledge and specific proprietary knowledge. The success of the new growth theory has pushed much work conducted in the economics of innovation tradition towards a research program more interested in the nature of the innovative firm, rather than the analysis of the system at large. Pages 25–35 provide an account of new research program centered upon the governance of the generation and distribution of technological knowledge and the analysis of the working of the knowledge commons. The following section on pages 34–35 elaborates the needs for new developments of the analysis of technological change as a systemic process which takes place in out-of-equilibrium conditions. This section and the conclusions also highlight the microeconomic limitations of much contemporary economics of innovation, the risks of an excess of attention to the entrepreneurial aspects of the innovating firm, and the need to develop further a broader framework of analysis, able to integrate the understanding of the determinants and the effects of both technological and structural change.

The discovery of the residual: the origins of the economics of innovation

The earliest definition of economics of innovation can be traced back to the 1950s with the introduction of the concept of residual. Thanks to the contributions of Abramovitz (1956) and Solow (1957) economics of innovation concentrates on the problem of explaining, at the firm level, the processes of growth of output which could not be easily attributed to an increase in production factors, under the constraints of equilibrium conditions, single and steady factors markets and constant returns to scale.

The residual emerges only when strong assumptions are made about increasing returns and equilibrium conditions. The removal of increasing returns is questionable and yet it has been in many ways a useful device. With increasing returns to scale output grows more than inputs and it is difficult to identify the specific contribution of technological change to economic growth. With constant returns to scale all increase in output which cannot be explained by means of appropriate changes in production factors can be considered as the result of the introduction of innovations in processes, products and in organizations. In out-of-equilibrium conditions the identification of the correct relative prices for both products and factors is all the more difficult.

The notion of residual risks overestimating the real role of technological change: part of the unexplained growth might be simply caused by increasing returns. Second and more important, this distinction and the actual separation between technological change and increasing returns risks hiding the role of increasing returns in the introduction of technological change and in turn the role of technological change in increasing returns.

Similarly, the analytical limitations of the removal of the recursive interactions between out-of-equilibrium conditions and the introduction of innovations are all the more evident: it is difficult to accept the hypothesis that innovations are introduced without even temporary departures from equilibrium conditions. On the other hand, however, the exclusion of increasing returns and out-of-equilibrium conditions as working hypotheses has provided an appropriate setting in which to appraise empirical evidence which highlights the key role of innovation in understanding both the economy and economics.

In this highly contextual framework, total factor productivity growth is a real puzzle for economics at large and especially for microeconomics. At the firm level the introduction of innovations cannot be considered as the outcome of standard rational behavior where marginal costs equal marginal revenues. When the costs of introducing an innovation match the revenue, measured either in terms of an increase in sales or a reduction in costs, no total factor productivity growth would take place (Griliches 1997).

The real challenge for economics of innovation is to provide an economic context to understand the behavior of economic agents facing radical uncertainty and multiple possible outcomes of their choices. A broader notion of rationality is needed as well as a more articulated understanding of the complexity of social interactions, beyond the standard price-quantities adjustments selected in a context of perfect foresight (Marchionatti 1999).

More generally innovation and total factor productivity growth in fact raise a serious problem for economics at large. The results of Solow (1957) show that more than 40 percent of the growth of the US economy in the years 1900–49 was determined by a factor which cannot be fully understood and analyzed with the traditional categories of standard economics. Systematic empirical work by the World Bank confirms that for a large array of countries total factor productivity growth accounts for more than a third of growth of output in the second half of the twentieth century (Chenery and Syrquin, 1975; Chenery *et al.* 1986).

The birth of economics of innovation as a specific area of enquiry and investigation in the broader context of the increasing specialization of economics can be considered the ultimate result of the analysis of growth of output and labor productivity, *ceteris paribus* input levels, as if increasing returns do not take place.

Manna, mines and technological opportunities

The manna metaphor draws mainly on the path-breaking contributions of the economics of technical change. At the beginning of the 1960s the hypothesis that technical progress was intrinsically exogenous is put forward as a methodological device to better understand its effects on the economic system. The assumptions about full exogeneity had the advantage of making it possible to disentangle the analysis of technological change from the complex web of other dynamic forces and their interplay.

Specifically in the manna tradition of analysis a linear sequence between scientific discoveries and technological innovations is introduced. Scientists deliver inventions and new scientific knowledge. Scientific knowledge eventually translates into technological knowledge which in turn feeds the introduction of technological innovations (Machlup 1962).

Since the first introduction the manna metaphor relied heavily, as an external reference, on the first sociology of innovation. Sociology of innovation drew mainly on the early contributions of Merton, in turn based upon the Weberian tradition, in the attempt to identify the objectives and incentives of scientific undertaking. Scientists, mainly academics, generate new scientific knowledge in an appropriate institutional context, one where incentives are not defined in strict economic terms. Scientists generate scientific discoveries in the form of public science in order to achieve peer reputation. Publications increase the stock of knowledge available on the shelf and ready to be used, for economic purposes, by firms.

The seminal contributions of Kenneth Arrow (1962a) has long shaped the debate about economic organization for the supply of technological knowledge and hence technological innovations. In this approach technological knowledge was seen as a public good for the high levels of indivisibility, non-excludability non-tradability and hence non-appropriability. In this context markets are not able to provide the appropriate levels of knowledge because of both the lack of incentives and the opportunities for implementing the division of labor and hence achieving adequate levels of specialization. The public provision of technological knowledge, and especially scientific knowledge, has been long regarded as the basic remedy to under-provision.

The public provision of scientific and technological knowledge by means of funding to universities and other public research bodies, as well as directly to firms willing to undertake research programs of general interest, found in this argument a rationale. This lead to the actual build-up and the systematic implementation of public knowledge commons.

Attributing the characteristics of a public good to scientific knowledge and therefore assuming indivisibility, non-excludability and non-appropriability, sanctioned a division of labor between firms and universities. The latter were of course responsible for the production and distribution of this public good. Firms had to be able to collect the stimulus, which was set off by new scientific discoveries. The state's role in this situation was that of an indispensable intermediary, which collected the taxes necessary to finance university research. Scientific inventions perfected and improved in an academic environment, and therefore, meta-economics, produced effects in terms of technological opportunities. Firms took these opportunities and introduced the innovations, thanks to which the total factor productivity increased and with it the amount of the income produced but not directly "explained" by the increase in inputs (Arrow 1962b and 1969).

In order to increase the incentives for innovators, the low levels of natural appropriability of technological innovations could be enhanced by means of intellectual property rights such as patents. Once more a dichotomy takes place. Science is public, while technology is private. Scientific knowledge is the primary source of technological knowledge. The former should remain in the public domain while the latter can be privatized in order to increase the rates of introduction of technological innovations (Lamberton 1971). Patents and copy-rights, if properly implemented and enforced, can reduce non-excludability and non-appropriability. In such an institutional design, intellectual property rights may also favor tradability and hence lead to higher levels of specialization and division of labor. Intellectual property rights can help increase the incentives for the production of scientific and technological knowledge.

Entrepreneurship in this context supplies evidence to the key role of meta-economic factors in assessing the rate and direction of technological change. Following Schumpeter Mark 1 – as Freeman *et al.* (1982) term the literature inspired by *The Theory of Economic Development* – the supply of entrepreneurs able to spot new technological opportunities and to understand the possible technological and economic applications of new scientific breakthrough is considered an important factor in understanding the pace of introduction of new technologies and their specific economic and technological characteristics. The analysis of the institutional and economic conditions which favor entrepreneurship and the entry of new innovative firms in the market place at large becomes an important area of investigation. The linkages and the interfaces between universities and new firms, the role of financial markets in the provision of funds to the innovations introduced by newcomers and the role of spatial proximity in fostering the natality of new high-tech firms all become object of systematic investigation.

The analysis of technological change as the result of an exogenous process provides the framework to study the asymmetric effects of the introduction of new technologies. New technologies can affect the marginal productivity of production factors and hence the derived demand for inputs when they are non-neutral and exhibit either capital-saving or labor-saving effects. The direction of technological change has important consequences for the equilibrium levels of the market price for inputs (Stoneman 1983).

In this context the analysis of the effects of technological change at the firm level is also relevant. New innovative firms enter the market place and destroy barriers to entry and monopolistic rents with the reduction of concentration. New technologies can be either centripetal or centrifugal according to their impact on regional and industrial concentration. In the latter case new technologies, such as electric power and lately advanced telecommunications, favor the even distribution of firms and plants in regional space and the reduction in their minimum efficient size. In the latter, new technologies, such as petrochemistry and assembly lines, may favor concentration when they lead to relevant economies of scale and scope.

An interesting shift takes place here: the introduction of technological knowledge and technological innovations is more and more viewed as the result of a process of mining without the help of geological maps. In this context, the notion of technological opportunities emerges as an important contribution. Firms both large and small, and industries, with high and low concentration levels, are more productive in terms of rates of introduction of technological innovations when technological opportunities are available. Technological opportunities are defined in terms of technical and economic fertility with respect to the efforts that are necessary for the introduction of new technologies and are conditional to scientific discoveries (Rosenberg 1976).

The strong evidence on the relevant asymmetries induced by technological change contrasts the assumption of full and even exogeneity of scientific break-throughs and technological innovations portrayed in the metaphor of the manna. The distribution of manna over the economic system seems far more complex and in any event uneven. The emergence of new technological opportunities and the size and extent of the new mines of technological knowledge are more and more seen as major factors affecting the performance of the system.

The classical hypothesis about the inducement of technological change by changes in factor prices receives new attention. Technological change is viewed as an augmented substitution. Firms are induced to change both techniques and technology by changes in the prices of production factors which alter significantly the basic characteristics of the production process. Technological change is now partly viewed as the outcome of the changes in the vertical relations among industries and firms in intermediary markets. All changes in the mix of production factors as determined by relevant shifts in their prices may induce the introduction of new technologies (Binswanger and Ruttan 1978).[1]

Similarly, an important contribution favoring a more endogenous presentation of technological change comes from macroeconomic studies which draw on Kaldor and the post-Kaldorian literature. The effects of demand growth on the introduction of new technologies are relevant to explain both the diffusion and the generation of new technologies. Embodied technological innovations can be adopted only when investments take place. While replacement investment is delayed by sunk costs, capacity investment can embody swiftly new technologies (Salter 1960). Evidence is also gathered on the faster flows of innovation in sectors in which the growth of demand is higher. This approach suggests that the economic system is able to condition, through the mechanism of expectations, scientific activity and to push its application in technological fields and industrial sectors where the expectations of demand growth, investment and hence profit are highest (Schmookler 1966).

Even though the generation of new knowledge remains an exogenous factor, its effective application in new technological innovations assumes strong endogenous connotations. The metaphor has proved useful as a focusing device and it leaves room for more articulated economic concepts which stress the asymmetric effects of the impact of the exogenous introduction of new technologies.

Although technological change is partly induced by market forces – with given and exogenous technological opportunities – the basic puzzle raised by total factor productivity growth cannot be solved. In a world of full "Olympic" rationality non-myopic agents should be fully aware of the possible effects of new technologies and search for them even without any pressure exerted by changes in factor prices or demand pull.

When an international perspective is assumed, the limits of the exogenous approach to understanding the origin of technological change is especially clear: countries differ widely in terms of both their ability to take advantage of new technologies, but also in their capability to innovate. The attempt to understand why technological change differs among and even within countries, not to say among and between industries and firms, promotes the application of biological grafting.

Biological grafts: epidemics and life cycles

Biology provides important suggestions and stimulation to the early economics of innovation. The identification and the analysis of recurrent regularities after the introduction of technological innovations supplies a large empirical evidence on which biological grafts provide significant framing opportunities.

A first relevant basket of important research programs favored by biological grafting is the analysis of the delays in the adoption of given technological innovations. The economics of the diffusion of new technologies is conceived as the study of the factors which account for the distribution over time of the adoption of identifiable successful innovations. A new technology is introduced after a scientific breakthrough and yet it takes time for all perspective users to adopt it. The successful and still widening application of the epidemic methodology emerges in this context (Griliches 1956).

Diffusion, like contagion, takes place in a population of heterogeneous agents which differ mainly in terms of information costs. Scientific breakthrough and in sequence technological innovations do fall from heaven, but they do have asymmetric effects in terms of the timing of adoptions. Within the same context – respectively the same industry, the same region, the same country – some firms adopt faster than others. Some innovations diffuse faster than others. For the same innovation adoption delays are longer in some industries, regions and countries than in others (Stoneman 1976, 1983, 1987).

Similarly, the analysis of diffusion faces significant changes when the analysis of the diffusion on the demand side is extended and applied to the diffusion on the supply side and the effects of the interactions are considered. Metcalfe (1981) provides a path-breaking context of analysis of diffusion where changes both in demand and in supply account for the distribution over time of adoptions. Metcalfe reintroduces the basic laws of standard economics into the epidemic framework and shows the relevance of their dynamic interplay.

The time distribution of adoptions can be conceived as the result of the spread of the contagious information about the profitability of the new technology. Proximity in geographical, industrial and technical space matters here in that it provides reluctant and skeptic, risk-adverse adopters the opportunity to assess the actual profitability of the new technology and hence to adopt it. The grafting of the epidemic analysis into economics of innovation can take place when contagion is assimilated to the spread of information. The key to understanding the vital role of the limits and constraints to perfect information raised by the information impact-edness of new technologies and the heavy requirements for Olympic rationality to apply in a context where new products and new processes are being introduced and the basket of goods upon which agents should comparative assessments is continually changed, are well disseminated in this context.

The second relevant biological graft onto economics of innovation is provided the life cycle metaphor. Ever since the Marshallian forest's trees, the life cycle metaphor has been around in the theory of the firm. A shift takes place when the sequence of natality, adolescence, maturity and obsolescence is applied to framing the steps in the life of a new product instead of a new firm. After introduction, the life of new product is characterized by a number of systematic events. According to the product life cycle approach a consistent pattern can be identified in the typology of innovations being introduced, in the evolution of the demand, in the industrial dynamics and in the characteristics of the growth of the firm.

The distinction between major innovations and minor ones is articulated and a sequence is identified between the introduction of a major innovation and the eventual swarm of minor, incremental ones. Second, a sequence between product and process innovations is identified. After the introduction of a new product, much research takes place in the effort to improve the production process (Abernathy and Utterback 1978; Utterback 1994).

Successful product innovations are eventually imitated by other firms. The swarm of minor incremental innovations is also and mainly the result of the imitative entry of new competitors based upon product differentiation. A sequence between early monopolies, long-lasting oligopolies and eventual monopolistic competition can be easily identified. Price costs margins decline with entry and imitation. Demand, however, also increases because of the epidemic diffusion of the information about the quality and the performances of the new product. The epidemic shift of the demand curves can match the shift of the supply curves engendered by entry rates. Incumbents can take advantage of the fast increase of demand and grow faster than newcomers. The analysis of the interplay between diffusion on the demand side and imitation of the supply sides remains an important cornerstone of the life cycle approach.

Especially when economies of scale matter, incumbents can enjoy the advantages of barriers to entry. The introduction of process innovations, following product innovations, often takes place in this phase, when incumbents try and react to the entry of new competitors. An important element of intentionality and causality finds here a first application. (Scherer 1984 and 1992).

The time patterns of entry of firms and hence the evolution of industrial demography, concentration, profitability and rates of growth of both firms and industries can be analyzed with the life cycle approach and significant regularities can be found and framed into a dynamic and coherent context.

The analysis of the role of small firms in the innovation process plays an important role in this context. A debate takes place between two approaches. The first relates directly to the so-called Schumpeterian hypothesis articulated by Schumpeter in *Capitalism, socialism and democracy* and articulates the hypothesis that large firms are necessary for high rates of technological advance to take place. Barriers to entry and monopolistic competition provide corporations with *ex ante* appropriability, reducing the risks of leakage and imitation. In turn large price-cost margins for corporations provide sufficient internal financial markets and hence competent decision-making with the liquidity and information necessary to fund new promising research activities. The second approach instead praises the role of new firms as vectors of new technologies and suggests that only high natality levels can sustain the rates of technological change. Much empirical evidence has been provided to support both hypotheses and the role of age as a factor shaping the complementarity of small and large firms, within the product life cycle, has been finally articulated (Acs and Audretsch 1990; Audretsch 1995; Audretsch and Klepper 2000).

At the firm level the product life cycle approach provides a theory of growth by diversification in products and nations. In fact it makes it possible to identify a sequence of introduction of complementary innovations when attention is paid to the technical and commercial proximity among innovations sequentially introduced. In this case firms can avoid the downturns associated with the obsolescence of an old product innovation by the introduction of a new product. The multinational growth of companies would also be the result of the product life cycle. Firms become multinational as they try and reproduce in different countries, ordered along a sequential gap in terms of revenue and demand characteristics, the monopolistic advantages associated with the introduction of a product innovation in each new country.

Relevant implications for international economics have been developed. In the first steps of the product life cycle, the new products are manufactured and consumed in the home country. After maturity and the multinational growth of firms, the products are manufactured and consumed abroad in the country of destination of the foreign direct investments. Foreign direct investment substitute for export. With obsolescence flows of exports from host countries to home countries take place. Foreign direct investments and imports are complementary.

The biological grafts have made it possible important progress. The understanding of the genuine sequential character of technological change, and of the out-of-equilibrium conditions, on both the demand and the supply side, in which innovations occur, has clearly benefited from the use of biological suggestions. The analysis, however, retains an intrinsic descriptive character and the main result is the identification of taxonomies and classification of sequential events (Pavitt 1984).

The study of the diffusion of new products on the demand side and of industrial dynamics on the supply side, developed in this context, retains the basic characters of an analysis of an adjustment process after an exogenous shock. The introduction of each new product and the generation of new technologies remain beyond the scope of the analysis. Their effects, however, are now better known: the equilibrium conditions of the system are perturbed, but eventually re-attained when, with the obsolescence of the product and the saturation of the diffusion, perfect competition on the supply side and perfect knowledge about the quality of the products on the demand side are restored.

Trajectories and technological paths

Three important contributions drawn directly from the philosophy of science and early cognitive science characterize the emergence of technological trajectories as the new heuristic metaphor and a research agenda. The distinction introduced by Polanyi (1958 and 1966) between tacit and codified knowledge, the analyses on the limits to Olympic rationality articulated by Simon (1947 and 1969) and the final entry of the notion of learning into the economic understanding of the characteristics of the economic man and the firm (Penrose 1959; Arrow 1962b) can be considered the founding blocks.

According to the path-breaking analysis of Polanyi, (economic) agents often know more than they are able to spell in a codified and explicit way. Tacit knowledge is embedded in the idiosyncratic procedures and habits elaborated by each agent. It can be partly translated into codified knowledge only by means of systematic and explicit efforts. The further distinction between tacit, artculable and codified knowledge, lately introduced by Cowan *et al.* (2000) stresses the complementarity between these levels of knowledge: not all the tacit knowledge can be transformed into fully condified knowledge.

Similarly, Simon introduces the notion of bounded rationality and elaborates on the distinction between substantive and procedural rationality. Agents cannot achieve substantive rationality for the burden of the wide range of activities necessary to gather and process all the relevant information. Agents can elaborate procedures to evaluate at each point in time and space the possible outcomes of their behavior, but within the boundaries of a limited knowledge. To cope with the constraint of bounded rationality, agents elaborate routines which make it possible to save on information costs and rely upon satisfying criteria as opposed to maximization.

The third key contribution to the new thinking is the notion of learning. Agents, and more specifically firms, are characterized by bounded rationality and limited knowledge: they are not able to fully articulate their knowledge, but agents as well as firm can learn. Learning is the result of repeated actions over time and reflective thinking. Learning has strong cumulative features and as such leads to dynamic increasing returns where cost reduction is associated with time rather than sheer size of production. Learning is also positively affected by feedbacks: learning is easier and more effective when it generates positive outcomes. Finally learning within organizations and learning by individuals have different characteristics. Learning of organization has effects, both on economics and the economy, that are even more relevant than learning by individuals (Penrose 1959; Arrow 1962b; Elster 1983; Dosi 1988; Malerba 1992).

New technologies may arise from such learning processes and especially from the efforts to convert tacit knowledge into new procedures which can be shared and transferred. A bottom-up understanding of the discovery process which complements the traditional top-down approach to the origin of technological innovations is now acquired (Kline and Rosenberg 1986).

Since the beginning of the 1980s, the manna metaphor was contrasted with the new metaphor of trajectories. The trajectory metaphor is first elaborated and introduced into economics by Nelson and Winter (1982). It assumes, first, the separation of technological knowledge from scientific knowledge and emphasizes the different rhythms of evolution of the two kinds of knowledge. The trajectory metaphor, second, emphasizes which elements contribute to the process of accumulating technological knowledge along both technical and behavioral axes of evolution. Demand pull and technology push are the driving engines which feed each other along such well-defined sequences of technological innovations. Finally and most important, building upon Simon and Polanyi, Nelson and Winter elaborate the notion of local search: firms search for new technologies in a technical space defined in terms of proximity with the techniques in use.

The notion of technological trajectory builds upon the achievements gathered with the product life cycle and makes possible a spring of cumulative research in the discipline. Rosenberg contributes the analysis of technological change with the notion of technological convergence which stresses the dynamic blending of technologies and their generative relations. The introduction of key technologies can activate an array of derivative innovations, based upon incremental technological changes (Rosenberg 1976 and 1982). The role of learning as main dynamic factor in the accumulation of tacit knowledge and eventually technological know-how receives new attention and appreciation (Arrow 1962b). The intrinsic cumulability of technological change along the trajectories is itself a major acquisition (Dosi *et al.* 1988). Incumbent innovators can take advantage of previous innovations in many ways: early competitive advantage makes it possible to fund new research; competence and technological knowledge acquired are useful inputs for further innovations; barriers to entry based upon market shares and size delay imitation; technological advance feeds diversification and entry in new industries.

An important reference in the trajectory metaphor is certainly Kuhn's contribution with his distinction between normal science and paradigmatic change. The trajectories then appear as a translation and application of the notion of normal science which moves along predetermined routes until its heuristic ability has run out. In such circumstances there are conditions for a solution of defined continuity, that is to say, paradigmatic change. Radical changes in the market place and in technology may induce firms to change their routines and introduce radical technological changes. Dosi (1982 and 1988) contributes to the analysis of trajectories with the notion of paradigmatic shift.

Paradigmatic crisis arise as factors of discontinuity. New trajectories are generated and old ones decline. The origin of such changes in technological paradigms however remains unclear but for the implicit reference to the notion of technological opportunities and their eventual exhaustion. The ultimate origin of technological change remains exogenous and a strong deterministic character is now added.

In this context, however, important contributions come from the analysis of the effects of the cumulability and tacitness of technological change. Trajectories are applied both to understand the dynamics of innovation with respect to the sequence of well-defined technologies and to the sequence of innovations introduced by well-identified firms and possibly economic systems, such as regions, industries and even countries.

The analysis of trajectories appears especially promising at the firm level and in the analysis of the competitive process. First-comers can now reap substantial competitive advantages and build barriers to entry based on their technological knowledge. An important implication of the distinction between tacit and codified knowledge consists in fact in the increase in the "natural" appropriability of technological knowledge.

The wealth of definitions in the analysis of the innovative process, the effort to integrate the analysis of innovative strategies with the study of the behavior of the firms in markets characterized by transient monopolistic competition and especially the emphasis placed on learning processes and on the accumulation of technological knowledge and economic competence remain important acquisitions, perhaps indispensable to the economics of innovation and new technologies. This rich understanding of the dynamics of technological change provides a context in which it is clear that the origin of innovation cannot be reduced to the output of research and development activities. This result contrasts with the emphasis on the analysis of the role of oligopolistic rivalry as the key inducement mechanism in the introduction of new technologies. Oligopolistic rivalry may provide a framework to understand the "equilibrium" amount of research and development expenditures but fails both to understand how research translate into innovation and consequently how total factor productivity grows (Dasgupta and Stiglitz 1980).

The trajectory metaphor appears to be a particularly fertile field of study of the behavior of large firms which are protagonists in growth and innovation strategies of an incremental kind in markets characterized by oligopolistic rivalries and high levels of product differentiation. In this literature, the large firm takes on a central role and appears, first if not exclusively, as the locus of accumulation of sticky technical knowledge and hence technological progress (Pavitt *et al.* 1989 and Pavitt 2000).

The biological notion of the selection process helps one grasp the sequential features of industrial dynamics along the trajectory. In this context, as well as the traditional sequence between monopoly, oligopoly and eventual monopolistic competition, an alternative route can be identified when the notion of dominant design is introduced. After the introduction of a variety of rival technologies by many rival firms, a selection process takes place and a few leading firms, able to elaborate the dominant design, emerge out of the competition with a consistent competitive advantage (Malerba 1996). Monopolistic rents emerge at the end of the selection process and may last. Adjustment to the exogenous shock of the introduction of a new technology – suggested by the product life cycle approach – never takes place and the system remains out of equilibrium.

An important contribution in this context is provided by the systematic application of the replicator elaborated by Metcalfe (1997). Metcalfe has shown the fertility of the replicator, a methodology originally conceived in biology to analyze the dynamics of species, in understanding the competitive process. Metcalfe applies the replicator analysis to show how innovators can earn extra profits, fund their growth and acquire larger market shares. The analysis of the diffusion of innovation is intertwined with the study of the selection mechanism in the market place. Firms that have been able to introduce new technologies are also able to increase their growth and their market shares.

Empirical research makes it possible to identify a variety of trajectories. The variety of trajectories emerges as serious limitation to its heuristic power. When a variety of trajectories in a variety of technologies and firms is identified such basic questions arise: why some trajectories are "steeper" than others; why some trajectories "last" longer than others; why some firms fail to innovate; why some industries are less able than others to build their own trajectory. Comparative studies across firms and regions facing similar technologies reveal the limitations of the trajectory in handling the cases of failure. Agreement on the trajectory metaphor disappears quite rapidly when its strong deterministic bent is fully revealed. This trajectory metaphor seems to revive the old temptation to use *ad hoc* technological determinism to explain the social and economic changes as a process of alignment of changes dictated by technology (Misa 1995).

The notion of bounded rationality introduced by Herbert Simon marks a major contribution to the economics of innovation. Olympic rationality is at odds with a context characterized by radical uncertainty where nobody actually knows the outcome of a research project and even less so the next directions of technological changes being introduced. As a matter of fact the very notions of future prices and future markets cannot even be considered when technological change is taken into account. The application of the notion of bounded rationality to economics of innovation leads to a new understanding of the basic inducement of innovation. Firms innovate when facing changes in the expected states of the world as generated by changes in both product and factors markets. Innovation is induced by the mismatch between unexpected events which myopic agents cannot fully anticipate and the irreversible decisions which need to be taken at any point in time. In this context the introduction of innovations and new technologies is the result of a local search, constrained by the limitations of firms in exploring a wide range of technological options. Procedural rationality pushes firms to limit the search for new technologies in the proximity of techniques already in use, upon which learning by doing and learning by using have increased the stock of competence and tacit knowledge (Nelson and Winter 1982; Antonelli 1995).

The empirical work conducted with the metaphor of the trajectory leads to more articulated economic concepts. In this context, characterized by the decline of the heuristic power of the notion of trajectory, increasing attention is paid to the role of historic time. The evidence provided by economic historians and historians of technology makes clear the key role of technological cumulability and irreversibility, and localized externalities.

The notions of localized technological change, introduced as early as in 1969 by Atkinson and Stiglitz and path-dependence, introduced by David (1975, 1985, 1987 and 1997a), receive new attention. Technological change is introduced locally by firms able to learn about the specific techniques in place and hence to improve them. Technological paths emerge as corridors in which firms are able to innovate, increase total factor productivity in a limited technical space so as to retain the original production mix. The notion of technological path substitutes the trajectory as the new heuristic metaphor and paves the way to an array of applications (Antonelli 1995 and 1999a).

At the microeconomic level the rate and direction of technological change can now be viewed as the endogenous outcome of the innovative sequential reaction of myopic firms induced by the interplay between irreversibility of their capital stock and disequilibrium in the products and factors markets which myopic agents cannot fully anticipate. Irreversible and clay capital stock can be thought to be constituted by both fixed physical capital and competence and technological knowledge in well-defined and circumscribed technical fields. Disequilibrium is relevant in terms of the variance in factors and products markets, of prices, costs and levels of demand, with respect to expectations and the related commitments in terms of fixed and irreversible production factors. Bounded rationality and limited knowledge affect the choice of firms which are unable to foresee all the possible events in the future, including the effects of the introduction of innovations, and must cope with the continual emergence of new events both in products and factors markets.

When irreversibility matters all changes in current business require some adjustment costs that are to be accounted for. In this approach firms are portrayed as agents whose behavior is constrained by the irreversible and clay character of a substantial portion of their material and immaterial capital. The conduct of firms moreover is affected by bounded rationality which implies strong limits in their capability to search and elaborate information about markets, techniques and technology. The introduction of technological changes is induced by the divergence between expectations and facts. Myopic agents are induced to innovate and introduce technological changes when the current state of affairs seems inappropriate and unexpected events take place. Myopic firms become aware of the costs of technological resilience. The costs of technological inertia are then confronted with the costs of innovation (Antonelli 1999a).

In such a context all changes in market demand and in the relative price of production factors are coped with by firms only after some dedicated resources have been applied to search for a new and more convenient routine. Consequently in this approach firms make sequential and yet myopic choices reacting to a sequence of "unexpected changes" in their business environment which are brought about by the introduction of innovation by other agents in both products and factors markets.[2]

Two notions seem relevant in this context: switching costs and localized knowledge. Switching costs are important when, because of the specific limitations imposed by the characteristics of the production process, firms find it difficult to change the levels of superfixed inputs they use currently. In these circumstances short-term production and cost theory apply to a longer time span than usually assumed. The raising portion of the standard short-term average costs curve defines the amount of switching costs firms exposed to unexpected changes in their production process must bear. Changes in the production mix and size of output expose firms to relevant price and output "Farrell" inefficiency, with a sharp decline in the efficient use of production factors.

Myopic firms are now induced to cope with the dynamics of demand and factor prices by introducing technological innovations and make the adjustments to market fluctuations yet retaining, as much as possible, the previous levels of superfixed inputs and hence change locally the technology, according to the relative costs of introducing innovations.

The introduction of technological changes in fact is not free nor is it the result of an autonomous process. It requires the intentional investment of dedicated resources to conduct research and development activities, to acquire external knowledge and take advantage of new technological opportunities and to elaborate upon tacit learning processes, to generate localized knowledge. Localized knowledge consists in fact of the accumulation of the benefits of experience and learning by doing, learning by using, learning by interacting with consumers, learning by purchasing. Firms are able to upgrade the existing technology only when they can blend the generic knowledge made available by new scientific discoveries and general movements of the scientific frontier with their own technological know-how. Hence localized knowledge makes it possible to capitalize on the generic knowledge with the know-how acquired using the techniques currently in use. This dynamic leads firms to prefer to remain in a region of techniques which are close to the original one and continue to improve the technology in use.

Firms induced to innovate by irreversibility and disequilibrium in both products and factors markets search locally for new technologies. The direction of tech-nological change is influenced by the search for new technologies that are complementary with the existing ones. The rate of technological change in turn is influenced by the relative efficiency of the search for new technologies. This is all the more plausible when the introduction of technological changes is made possible by the accumulation of competence and localized knowledge within the firm.

The notion of technological path-dependence is the result of the blending of the localized technological change approach and the broader issue of path-dependence. When path-dependence matters, economic action in each step is stochastically influenced by the past, but not deterministically caused by previous events. In a technological path the probability of introduction of each new technology is contingent upon previous innovations as well as cumulated technological competence, but also on the necessary complementarity of other factors such as the levels of irreversibility of the capital stock and the access conditions to local knowledge externalities.

An important distinction emerges here between internal and external path-dependence. Internal path-dependence takes place when the path along which the firm acts is determined by the irreversibility of its production factors. External path-dependence is determined by the external conditions. In the first case the choice of the new technology is influenced by the switching costs firms face when they try and change the levels of their inputs. In the latter case instead the choice of the new technology is shaped by the markets conditions. The notion of path dependence elaborated by Paul David (1975) belongs to the first case: firms are induced to follow a path of technological change by their internal characteristics. The notion of path dependence elaborated by Brian Arthur (1989) and Paul David (1985) clearly belongs to the second case: new technologies are sorted out by increasing returns to adoption at the system level.

Local externalities, as opposed to general ones, play a key role in this context. They are made available to innovators by the location in a conducive regional and industrial environment characterized by qualified interactions with other complementary innovators both upstream and downstream. In so doing external technological path-dependence is reinforced by the role of complementarities and interdependence in the generation of new technological change and in the introduction of new technologies. Increasing returns based upon network externalities on the supply side in the production of new technologies add to increasing returns to adoption based upon network externalities on the demand side.

The notion of path-dependence seems more able than trajectories to accommodate both success and failures and to explain the variety of possible outcomes. The fragile mix of complementary and yet necessary conditions affecting the transition from each step to the next along the technological path becomes key to understanding the actual sequence of events in assessing the rate and the direction of technological change.

A strong common thread links the analyses developed with the notion of technological trajectory and the notion of path-dependence: technological change takes place in out-of-equilibrium conditions and it is fed by out-of-equilibrium conditions. From this viewpoint the technological trajectory and eventually the technological path represent a significant progress with respect to the product life cycle. In the latter approach in fact the analysis focused the adjustment process, after the exogenous shock brought about the introduction of an innovation, as a return to equilibrium conditions. In this new context, instead, the introduction of innovations takes place because of disequilibrium conditions of the system and reproduces new disequilibrium conditions.

Technological change is now the endogenous outcome of a disequilibrium condition which has little chance of converging towards a new equilibrium. Actually equilibrium and technological change emerge as opposite extremes: equilibrium is possible when no technological change takes place and vice versa. Technological change is possible only in market conditions that are far away from equilibrium.

Collective knowledge and networks

During the 1990s the spread of constructivism in the history and in the sociology of science (Latour 1987; Bijker 1987; Callon 1989; Smith and Marx 1995) and the emergence of the new science of complexity (Rycroft and Kash 1999) parallel the discovery of new prospects in the field of innovation. In the analysis of the conditions governing the generation and the distribution of new knowledge, the economic system is viewed as a structure of relations and interactions, including transactions, that are likely to shape the actual speed and rate of introduction of new technologies. A systemic approach to the analysis of the introduction and diffusion of new technologies is progressively implemented.

In this context, the revival of the Marshallian analysis of externalities, of various shades and of the Marshallian competitive process offers numerous interpretative points about the role of the institutional set-up into which the virtuous interactions among innovators and the actual flow of technological spillovers can take place. This analysis brings into the center of the discussion the basic issue of the intrinsic complementarity and interdependence, at the technological, industrial and regional levels, among the agents in the accumulation of new technological knowledge and economic competence and subsequently in the introduction and adoption of new technologies. The need of a systemic analysis able to integrate the understanding of interactions based on the price mechanism, viewed as the single vector of relevant information, with the analysis of the variety of interactions that make the knowledge spillover possible, becomes evident (Audretsch and Feldman 1996; Audretsch and Stephan 1996; Loasby 1999).

The generation and introduction of technological innovations are now viewed as the result of complex alliances and compromises among heterogeneous groups of agents. Agents are diverse because of the variety of competencies and localized kinds of knowledge they build upon. Alliances are based upon the valorization of weak knowledge indivisibilities and local complementarities among technological different kinds of knowledge. The convergence of the efforts of a variety of innovators, each of which has a specific and yet complementary technological base, can lead to the successful generation of a new technology. The issues of complementarity, weak divisibility and technological interdependence become central to an understanding of the attributes of specific technological, industrial and regional systems or networks in which the dissemination and access to technological knowledge takes place (Freeman 1991; Nelson 1993; David and Foray 1994).

Similarly, the analysis of the conditions for the appropriability of knowledge makes it possible to the economics of innovation to understand the key role of technological externalities in the generation of new knowledge and the positive effects of technological spillovers (Griliches 1992). The discovery of external knowledge, available not only by means of transactions in the markets for knowledge, but also by means of technological interactions, marks a new important step in the debate. External knowledge is an important input in the production process of new knowledge. The appreciation of external knowledge, as an essential input in the production of new knowledge, contributes to the articulation of the systems of innovation approach, where the production of knowledge is viewed as the result of the cooperative behavior of agents undertaking complementary research activities (Antonelli 2001).

The systemic characters of the generation and use of new knowledge are emphasized and stressed. Market transactions play only a partial role in the understanbding of the systemic generation of new knowledge. Knowledge interactions are most relevant and include a variety of actors and institutions. According to the specific form of interaction considered, a variety of systems of innovation has been identified: national, regional, industrial and technological. Each systemic dimension stresses the role of specific forms of complementarity and interdependence among actors in the generation of new knowledge and in its eventual use for the implementation of new technologies (Carlsson and Stankiewitz 1991; Carlsson 1995 and 1998; Edquist 1997).

The production of knowledge, both scientific and technological, now appears to be strongly conditioned by the social, institutional and economic conditions, in which it takes place (Gibbons *et al.* 1994). Interaction and communication among a variety of innovative agents and in the field of scientific production among scientists as well as between universities and companies plays a central role in the interpretation of the factors which are at the origin of the rates of production of new knowledge and of its specific direction, intended as specific fields of application (David and Foray 1994; David 1993 and 1994). The results of the empirical analyses of Lundvall (1985) and Von Hippel (1988) on the key role of user–producer interactions as basic engines for the accumulation of new technological knowledge and the eventual introduction of new technologies play a key role at this stage. The production of knowledge in fact, becomes a central condition for the subsequent perfection and improvement of technological innovation.

The costs of exclusion associated with intellectual property rights, as a consequence, should be taken into account. Monopolistic control of relevant bits of knowledge, provided both *ex ante* and *ex post* by patents and barriers to entry in the products markets respectively, can prevent not only its uncontrolled leakage and hence its dissemination but also further recombination, at least for a relevant stretch of time. The advantages of the intellectual property right regime, in terms of increased incentives to the market provision of technological knowledge, are now balanced by the costs in terms of delayed usage and incremental enrichment. The vertical and horizontal effects of indivisibility display their powerful effects in terms of cumulability. Indivisibility of knowledge translates into the basic cumulative complementarity among bits of knowledge. Complementarity and cumulability in turn imply that new bits of knowledge can be better introduced by building upon other bits already acquired, both in the same specific context and in other adjacent ones. The access exclusion from the knowledge already acquired reduces the prospect for new acquisitions and in any event has a strong social cost in terms of duplication expenses (David 1993).

The analysis of the process of accumulation of technological knowledge plays a key role in this context, one where historic time is more and more taken into account. Technological knowledge is now viewed as a quasi-public good: it can be partly appropriated by innovators and it spills over to third parties although less easily than is assumed in the public good tradition of analysis. The important role of tacit knowledge, embedded in the organization of innovators and especially in their learning procedures, reduces the capability of prospective imitators to absorb the new knowledge and favors higher levels of appropriability. Specific activities must be implemented in order to take advantage of the spillovers. The interaction between universities and research centers at large and the business community provide a new ground for much empirical analysis (Griliches 1992; David 1993 and 1994; Etzkowitz and Leydesdorff 2000; Foray 2000).

The amount of external technological knowledge available in a given context, either industrial, technological or regional, becomes an important endowment, as well the condition of access to it and the characteristics of the relational set-up. A variety of players contributes to the amount of external technological knowledge, e.g. firms, universities and research centers, as well as brokers and other undertakings specialized in the spread of technological knowledge such as knowledge-intensive business service activities. The institutions of labor markets play an important role: job-seniority and wage structures can modify the flows of technological knowledge especially in a regional context. Interindustrial division of labor and outsourcing in general also play an important role as they increase the flows of technological communication. Knowledge-intensive business service activities emerge as providers of technological knowledge and complementary actors in the trade of patents and other intellectual property rights.

The collective character of technological knowledge and the complementarity of limited areas of knowledge put at the service of each actor are emphasized here. On the specific level of the economic analysis the cooperative way that innovations are perfected and improved is highlighted. It is seen as a process which also involves rival firms and firms situated in the same technological districts.[3] Innovation is set off in such situations by the cross-fertilization of specific and yet complementary bits of knowledge and their continuous development. The range and the variety of the actors involved in these exchanges appear to be determinant factors. The increasing number of the channels of communication among heterogeneous subjects, bringing diverse knowledge, but no less susceptible, is key to activate new complementarities which call for the participation and the verbalization of experiences. The parallel development of new technologies of information and communication highlights this approach and emphasizes the fundamental role that interaction and communication play as factors of production of new knowledge through processes of formation and contamination. In this context the use of graph theory and the contamination of the percolation approach, originally conceived in physics, seem useful attempts to analyze the architecture and the role of the web of complementarities and interdependencies which relate firms and technologies and cannot be fully signaled by prices (David 1998; Antonelli 1999 and 2001a).

In this approach the distinction between innovation and diffusion is blurred and, on the other hand, each adoption is viewed as the result of a complementary effort that makes useful and specifically reliable a new technology, increasing its scope of application. Adopters are no longer viewed as passive and reluctant perspective users, but rather as ingenious screeners that assess the scope for complementarity and cumulability of each new technology with their own specific needs and contexts of action. Profitability of adoption is the result of a process rather than a given fact.

A technology diffuses when it applies to a variety of diverse conditions of use. The intrinsic heterogeneity of agents applies in fact not only to their own technological base but also to the product and factor markets in which they operate. The vintage structure of their fixed costs and both tangible and intangible capital can be portrayed as major factors of differentiation and identification of the specific context of action both with respect to technological change and market strategy.

New ideas can be implemented and incrementally enriched, so as to become eventually profitable innovations, only when appropriate coalitions of hetero-geneous firms are formed. The notion of probit-diffusion contrasts with the epidemic approach. New technologies are adopted only if and when they fit specific product and factor markets conditions: some agents will never adopt a new technology and the identification of the determinants of the non-adoption becomes relevant (Stoneman 1995).

The diffusion of a new technology is no longer seen as the outcome of the adaptive adoption of a new single technology, but rather of the choice of one new technology among many. Diffusion is the result of the selection of a dominant design out of an original variety of different technological options. Interdependence among users may lead to increasing returns in adoption so that technologies that have been adopted by a large share of perspective users have higher chances to win out the selection process and spread to the rest of the system (Arthur 1989 and 1994).

The analysis of industrial dynamics supplies the arena into which the market construction of new technologies can be observed. A variety of firms can be analyzed in the course of their competitive strategies where technological change plays a major role. Each firm is characterized by a specific competence upon which localized knowledge has been implemented. Each firm is also characterized by a specific endowment of fixed capital, and specific contexts of conduct in both factor and product markets. The changing market conditions induce firms to innovate both in market strategies and in technologies.

Technological choices concerning the introduction of product and process innovations, the adoption of new technologies provided by suppliers and the imitation of competitors is mingled with market strategies such as specialization, outsourcing, diversification, entry and exit, merger and acquisitions and internal growth. In a continual trial in the market place firms experiment with their changing mix of technological and market conduct. At the aggregate level the result is the market selection of new and better technologies, often characterized by strong systemic complementarities. The changing coalitions between different groups of players in overlapping and yet specific technological arenas shape the rate and direction of technological change at large.

Much advance in this context is provided by the new understanding of the dynamics of network externalities. Network externalities apply not only to the demand side when the utility of a given product is influenced by the number of users, but also on the supply side when the productivity of a capital good is influenced by the number of users. Moreover network externalities apply also to the generation of new knowledge and the introduction of new technological systems when the positive effects of the increasing number of complementary kinds of knowledge and related technologies are considered. The adoption of innovation, once introduced, is also affected by the number of early adopters (Katz and Shapiro 1985; Antonelli 1992 and 1999; Cohendet *et al.* 1998).

Network externalities apply moreover when each new technology is comple-mentary to other parallel innovations and contributes to the overall efficiency of the system. A technological system emerges when interdependent and complementary innovations are eventually introduced. The definition of general purpose technology introduced by Bresnahan and Trajtenberg (1995) and further elaborated by Lypsey *et al.* (1998) applies very well to new information technology. The systemic character of general purpose technologies is relevant both with respect to their origins and with respect to their applications. New information and communication technology, like previous general purpose technologies, is the result of the complementarity and interdependence of a variety of technological innovations being sequentially introduced. New information and communication technologies, like previous general purpose technologies, have high levels of fungibility as they apply to a great variety of production processes of the new technological system including a great variety of pre-existing products and processes.

As in the case of previous general purpose technologies, such as the railway, the dynamo and mass production technologies, new information and communication technologies are the result of a myriad of complementary innovations where each introduction has been the result of the complementarity and interoperability with other parallel innovations and has reinforced the scope of application and the productivity of the others. The process is such that no product and process can be manufactured without the substantial application of new information and communication technologies or without substantial effects of the application of new information and communication technologies. The case of general purpose technologies stresses the role of complementarities and interdependence both in the generation and use of new technological knowledge.

The increasing understanding of the systemic characteristics of processes leading to the introduction and selection of the new technologies pushes towards a new representation, one where technological change is viewed as a form of systemic, dynamic, stochastic and finite increasing returns which lead to punctuated growth. Technological change in fact takes place when a number of highly qualified necessary conditions apply. The successful introduction of technological change can be seen as the fragile result of a complex set of necessary and complementary conditions.

The appreciation and identification of the structural conditions which shape economic systems and are conducive to the introduction and diffusion of new technologies is the main result of this line of analysis. Changes in technology may take place and they may lead to the increase of output, with a given level of inputs, but only when a general process of economic change and industrial dynamics with well-defined necessary conditions at the system level is actually in place. In this context a large number of complementary and necessary conditions, both technological, as well as institutional, competitive and macroeconomic, at the system level, play a central role for a fast rate of introduction of new technologies to take place.

In this context Metcalfe provides new ground for the need to elaborate a systemic analysis of the behavior of economic agents which takes into account, next to the characteristics of factor and input markets, the specific features of technologies and technological knowledge, considered as specific articrafts and processes. Technological choices need to be analyzed in conjunction with other behaviors and choices in terms of prices and quantities. In such a systemic approach innovations and firms cannot be analyzed in isolation but only within a dynamic system of technological, industrial and regional interdependencies and complementarities, both on the supply and the demand side, which are not exclusively and fully mediated by the price mechanism (Metcalfe 1995 and 1997).

Dangerous currents

Success often breeds risks. The wealth of results achieved by the economics of innovation and the parallel evidence on the economic relevance of the new wave of radical innovations clustering around new information and communication technologies has solicited equilibrium analysis.

The build-up of an economics of intellectual property rights has made important contributions to this approach. The strong hypothesis that appropriate implementation of patents, finely tuned in terms of scope, duration, assignment and renewal procedures, can reduce or even erase the problems raised by the public good character of technological knowledge has been articulated.

At the same time much empirical evidence and theoretical research has shown that appropriability is *de facto* much higher than assumed. Knowledge is contextual and specific to the original conditions of accumulation and generation: as such natural appropriability conditions are far better than assumed. Imitation costs seem high as well as the costs of receptivity and re-engineering necessary to make use of non-proprietary knowledge. The costs of the 'non-invented-here-syndrome' are appreciated. The assistance of original knowledge holders to perspective users is relevant, if not necessary.

These two strands of analysis, *ex post*, contributed with complementary arguments to the new hypothesis that the supply and the demand for technological knowledge can be identified, the actual creation and implementation of markets for technological knowledge is possible, and the results of such market interactions are compatible with a workable competitive system in the proximity of equilibrium conditions.

An important effort takes place in order to integrate the analysis of technological change into an equilibrium context of analysis: the new growth theory. The new growth theory builds upon three important acquisitions of economics of innovation: (a) the distinction between generic and tacit knowledge, and the related notion of technological knowledge as a quasi-public good because of quasi-appropriability; (b) the understanding of technological externalities and the dynamics of spillover, and (c) the notion of monopolistic competition as a result of the introduction of new products.

The new growth theory draws basic inputs from the developments of the economics of innovation but elaborates an analytical framework where the accumulation of knowledge and the eventual introduction of innovations is explained in an equilibrium context.

According to Romer (1994) economic growth relies upon the collective access to generic knowledge which flows in the air. Romer distinguishes between generic technological knowledge, germane to a variety of uses, and specific technological knowledge embodied in products and as such with strong idiosyncratic features. Specific knowledge can be appropriated, generic knowledge instead retains the typical features of the Arrovian public good. Innovators generate generic knowledge while they are engaged in the introduction of new specific knowledge embodied in new products and new processes. The production of specific knowledge takes advantage of the collective availability of generic ones. The spillover of generic knowledge helps the generation of new specific knowledge by third parties and yet does not reduce the incentives to the generation of new knowledge for the strong appropriability of the specific applications. Monopolistic competition characterizes the markets for products and provides a coherent context for a close variety of products, drawing from the same pool of generic knowledge, to coexist.[4]

In this research program, however, the wealth of institutional analysis, keen to grasp the specific conditions into which the dynamics of increasing returns activated by technological externalities is available, is lost. According to the new growth theory technological externalities can spill everywhere and no specific conditions need to be identified.

This new approach led not only to endogenous growth theorizing but also to significant steps towards the privatization of public knowledge commons. Universities were solicited to patent their discoveries and often forced to enter the markets for the technological outsourcing of large corporations (Argyres and Liebeskind 1998). Public funding for research activities declined and they were questioned, if not put under strain. A closer look at the working of the public commons and the actual need to put under scrutiny the productivity of the resources invested in the public knowledge commons, both at the system and the single units level, was advocated (Geuna 1999). Some attempts to liberalize the markets were also made, especially in the new general purpose technology field of new information and communication technologies with the divestiture in telecommunications and a new more aggressive antitrust stance (David 1997a).

Similarly, the analysis of the disequilibrium conditions which are the necessary condition for innovation to take place is eradicated. Innovation is fully routinized and becomes part of daily management. The linear and smooth growth of the economic system is pushed by the continual introduction of innovations by firms which operate in a monopolistic competition context and are able to generate new specific knowledge, and it is fed by the free spillover of general technological knowledge but it is not affected by the interaction between technological change and structural change.

The governance of knowledge commons

The understanding of the actual governance conditions for the production and distribution of technological knowledge is a central issue. Technological knowledge can be generated and effectively distributed, in an economic system, only when a number of highly specific and idiosyncratic conditions are satisfied. Their identification and eventual implementation become a central target both for economic analysis and economic policy.

In resource-based theory the firm is viewed as the primary institution for the governance of knowledge commons and hence for the accumulation and generation of technological and organizational knowledge. The firm is first of all the institutional locus where technological and organizational knowledge is generated by means of the integration of learning processes and formal research and development activities. The firm is considered in this approach primarily as a depository and a generator of competence. Such competence applies to the manufacturing processes as well to the management of internal coordination and to the procedures and the skills that are necessary in order to use the markets (Penrose 1959; Foss 1997).

The resource-based theory of the firm has grown as a development and an application of the economics of learning. The enquiry about the dynamics and the characteristics of learning processes, such as learning by doing and learning by using, and their relevance in explaining technological change has led to the identification of the firm as the primary locus of the generation and valorization of knowledge immediately relevant for the economic action, at least in market economies (Loasby 1999).

The resource-based theory of the firm focuses the attention on the characteristics of the process of accumulation of competence, the generation of technological knowledge and the introduction of technological and organizational innovations, as key factors to understanding the firm. The characteristics of knowledge, such as appropriability, cumulability and complementarity, and its state, whether tacit, articulable or codified, play a major role in understanding the architectural design of the firm and the combination of activities retained within its borders. Parallel to knowledge, competence is a central ingredient in the resource-based theory of the firm. Competence is defined in terms of problem-solving capabilities and makes it possible for the firm not only to know-how, but also to know-where, to know-when, and to know what to produce, to sell, to buy, to coordinate and to innovate (Nooteboom 2000).

The notions of localized technological knowledge and localized technological change contribute to this approach as they stress the relevance of the learning processes circumscribed in the specific and idiosyncratic locations, within technical, organizational, product and geographical spaces, of each firm at each point in time. The learning processes in such locations are the basic conditions for the accumulation of experience and the eventual generation of both competence and tacit knowledge. On these bases in turn the firm is able to acquire other forms of knowledge, respectively external codified and tacit knowledge and to implement the internal tacit knowledge with research and development activities. In this approach, the firm is primarily defined as a bundle of activities that are complementary with respect to the generation of knowledge and competence (Antonelli 1999 and 2001a).

The resource-based theory of the firm has been much implemented in the managerial literature as well as in economics, and especially in evolutionary economics where it provides important contributions to its micro foundations (Teece 2000; Foss 1998). Economics of knowledge also contributes the resource-based approach. The intrinsic indivisibility of knowledge is assumed as axiomatic. Knowledge is indivisible and yet dispersed in a variety of individuals. No individual can claim the full control of all knowledge. Complementarity among individuals is central in the accumulation, generation and eventual valorization of knowledge. The firm is the institution which is better able to manage such complementarities in a market economy. Hence the firm is primarily a governance mechanism for the production of knowledge.

In the resource-based theory the firm is much more than a nexus of contracts. Yet important complementarities can be found with transaction costs economics when an effort is made to understand the role of competence and knowledge in the definition of the borders of the firm, under the constraint of the resources that are necessary to coordinate the diverse activities retained within its borders. The generation of knowledge is the primary role of the firm but under the constraint of the economics of governance.

The firm itself is more and more regarded as an island of coordination procedures that facilitate the accumulation of knowledge. The Coase-Williamson argument, much applied to the choice between coordination and transaction in the organization of the economic activity, can now be stretched and elaborated so as to understanding the fabric of technological knowledge (Furubotn 2001).

Within corporations the coordination of technological communication becomes a relevant issue. The organization of firms appears to be influenced also by the need to implement and valorize the complementarity of the bits of knowledge possessed and accumulated in the diverse units. The distinctive notion of the costs of technological transactions and interactions can also be identified. The trade-off between knowledge coordination costs and knowledge transaction and interaction costs contributes the understanding of the technological choices of the firm (Argyres 1995).

In the governance of knowledge not only the traditional "make or buy" trade-off is relevant, but also a "make or sell" choice has to be considered. The firm in fact needs to assess not only whether to rely upon external or internal knowledge in the production of new knowledge, but also whether to try and valorize the knowledge available internally as a good itself and sell it disembodied in the markets for technological knowledge, or to use it as an input in the production of other goods (Teece 1986).

A wide range of choices in terms of governance can be analyzed and understood with respect to the characteristics of the processes of knowledge generation and usage. Technological strategies can be implemented by means of internal research and development laboratories, technological outsourcing, location of research and development centers into technological districts, technological alliances and research joint-ventures and finally actual mergers and acquisition (Howells 1999; Antonelli and Quéré 2002).

The issues of the distribution of knowledge become central in the debate and the notion of an actual knowledge trade-off is articulated. Uncontrolled leakage and low appropriability regimes reduce incentives and lead to under-provision. Excess appropriability, both *ex ante* and *ex post*, however, may slow down if not impede the working of knowledge complementarity, cumulability and fungibility. A governance of the knowledge trade-off is necessary both at the firm and at the system levels (Mazzoleni and Nelson 1998; Kingston 2001).

Indivisibility is now perceived as an important attribute of knowledge as an economic good. High levels of complementarity and low levels of separability among units of knowledge, both in its generation and in its use, have major consequences. The costs of all limitations to the full circulation, rapid dissemination and sequential usage of knowledge become more and more evident. A trade-off now emerges between the advantages of the intellectual property rights (and *ex-ante* barriers to entry), in terms of increased incentives to the market provision of technological knowledge and the opportunity costs stemming from the missing opportunities for incremental advances and recombination both within each field and across fields (Kingston 2001).

The analysis of the governance of both the generation and usage of technological knowledge, that is of the mechanims designed and of the conditions of access and exclusion to the flows of technological interactions, transactions, coordination and communication that are specifically designed to handle the generation and the distribution of technological knowledge deserves careful assessment and scrutiny (Menard 2000; Carroll and Teece 1999; Williamson 1985 and 1996; Langlois 1986).

Inclusion needs to be coordinated and managed. Free-riding can take place, although reciprocity and mutuality in interactions based upon knowledge barters, implemented by repeated and long-lasting exchanges, can help reduce the extent and the effect. Exclusion is dangerous and risks jeopardizing the relevant complementarity which characterizes the generation of new technologies (Geuna *et al.* 2002).

The identification of the agents holding specific bits of knowledge and the assessment of their complementarity becomes an important function. This is expensive both in terms of search and opportunity costs: the costs of interacting with the wrong agents in terms of low opportunities. A specific form of knowledge transaction cost can be identified here. The selection of the firms and agents with whom technological cooperation and technological communication can take place is a relevant aspect of the governance mechanism and of the governance process. This may lead to the creation of technological clubs and research joint-ventures as institutional organizations designed to carry on collective research within selective coalitions.

The economics of technological knowledge has made important progress in the identification of specific characteristics of technological knowledge. The forms and the types of knowledge matter. Different governance mechanisms and governance choices emerge according to the characteristics of technological knowledge. The forms of the relevant technological knowledge matter: whether technological is more tacit, articulable or codified has a direct bearing on the governance of the accumulation process (Cowan *et al.* 2000).

The exchange of tacit scientific and technological knowledge seems easier within research communities because it is based upon repeated interactions and closed reciprocity in communication. Random inclusion can take place with positive effects, provided newcomers are properly selected. The incentives to the creation of informal interaction procedures, often implemented by co-localization within technological districts, are very strong in this case. Collective bodies emerge as important governance structures especially when technological knowledge is tacit and articulation requires complex procedures.

The exchanges and interactions in articulable knowledge occur more readily within technological clubs and coalitions where membership is closely assessed and selectivity requirements are far higher. The reputation of the fellows in the club plays an important role in building closed research coalitions (Teece 2000). When technological knowledge is more articulable, the contractual interaction among partners within research joint-ventures and technological clubs can be better implemented.

The distinction between procedural and content contracts is relevant here. Procedural contracts are incomplete with respect to the actual definition of the content of the exchanges, but are able to specify the modality and the sequence of the interactions. Content contracts better focus the characteristics of the actual transaction. Specific procedural contract can be designed and possibly enforced to deal with the process of participation, the sequence of obligations, the timing of involvement and of the assignment of property rights, temporary and partial exclusivity, time lags and partial domains of privilege to relevant contributors, depending on both the amount of inputs and the actual results.

In this context the interactions and the transactions between the business and the academic communities seem to find their specific context of implementation. The interface between tacit and codified knowledge, defined as articulable knowledge, provides in fact the opportunity to define the points of common interest between the two parties while the broader goals of the scientific undertaking in terms of the commitments towards the creation of intellectual commons can still be pursued.

Codified technological knowledge better meets the conditions for tradability especially if implemented by an appropriate intellectual property right regime and when the assistance of inventors and as such vendors to perspective users is necessary and useful to reduce adoption and adaptation costs. The markets for technological knowledge with actual arms' length transactions are often found in this context. The design of actual content contracts, such as in the case of licences, is possible and enforcement more reliable (Arora *et al.* 2001).

Next to the forms of technological knowledge, its types play an important role. The levels of complexity, fungibility, cumulability and tradability of each bit of technological knowledge can be fully appreciated when assessing the governance mode of the generation and usage of new technological knowledge.

The high fungibility of the technological knowledge generated by each firm, especially if associated with high levels of cumulability, provides important incentives towards internalization and hence diversification, both lateral, vertical and geographical. Diversification, integration and multinational growth can be seen as strategic choices elaborated by firms in order to valorize intangible assets which cannot be traded as such in the markets for disembodied knowledge. Growth thus is a tool used to valorize intangible assets which cannot be appropriated by means of intellectual property rights, but only when embodied in traditional property rights. The firm has in fact the opportunity to appropriate the rents stemming from the application of its knowledge in the production of previously unrelated goods. Qualified user–producer interactions are also useful when the application and valorization of fungibility requires the active involvement of downstream or upstream actors.

Diversification, however, can be the consequence of the generation of new knowledge, as well as a tool, hence a factor, in the knowledge generation process. External growth in fact can be the result of a strategy directed to implement the core competencies with the internalization of external ones. Growth in new business fields here is guided by the search for complementarity. Those which play a strategic role make technological innovations possible.

The greater the complexity of the technological knowledge necessary to generate new technologies, the more likely is the implementation of strategies based upon technological outsourcing. When the technological base of a firm is complex and requires the integration and recombination of a large variety of technological domains, the use of external knowledge is encouraged by the high levels of internal coordination costs of the diverse sources and competencies that are necessary.

Technological outsourcing can take place in many different ways. External knowledge can be accessed by means of actual transactions of patents and licences, the purchase of research and knowledge services from knowledge-intensive business services firms including universities and other research centers, the location in knowledge-intensive districts and finally the acquisition of knowledge-intensive firms. Financial markets can be seen as markets for knowledge where technological knowledge is no longer embodied in capital goods, intermediary inputs or skills, but directly in financial assets. The acquisition of a firm can be an effective form of accessing external knowledge

Conversely, the greater the cumulability of the technological knowledge, specific to the products and the production process of a firm, the larger are the incentives towards the internalization of the knowledge generation process. Technological outsourcing in fact has high costs in terms of missed opportunities for further advances. The same argument applies when learning plays a key role in the generation of new knowledge: the full control of the production process is likely to yield important benefits in terms of increased rates of accumulation of new technological knowledge.

The understanding of the governance of technological knowledge and of the demand-side externalities in technological knowledge makes possible important contributions to the economics of governance. Governance structures not only depend upon the characteristics of the transactions and of the production processes for given technologies and within the boundaries of the existing technological base: are also influenced by the role and the features of technological knowledge. The governance of the generation and usage of technological knowledge emerges an important area for empirical and theoretical investigation. The governance of the knowledge trade-off has many important implications not only at the firm level, but also with respect to the system at large (Teece 2000; Nelson and Sampat 2001).

The notion of knowledge fungibility plays an important role in this context. It seems clear that the larger the fungibility the wider is the scope of application and recombination of any specific bit of knowledge and the larger the costs of exclusion. General purpose technologies should be more accessible than specific single usage technologies. The assignment of intellectual property rights should be tuned, according to the social costs of exclusion from specific portions of technological knowledge, according to their scope of application and to their relevance with respect to further discoveries. The definition of the domains of cumulability becomes most relevant. Modularity seems pertinent also in this context. Chains of weak and strong complementarity and cumulability can be detected and modules of technological knowledge can be identified. The effects of utility interdependence can be mapped into well-defined regions with borders designed by the actual extent of knowledge complementarity and cumulability (Antonelli 2001a).

The identification of such modules in turn becomes relevant from a strategic viewpoint at the company level. Certain firms that are a depository of certain bits of knowledge are likely to be more interdependent than others with some subclasses of other agents. The externalities spilling from their own research agenda and their own cumulated competence can be more relevant than others'. The identification of technological modules and the drawing of specific knowledge maps into which each agent is placed can become a tool to activate and valorize the innovation capability of both the firm and the system with proper policy strategies.

The understanding of the actual levels of cumulability, fungibility and complementarity of well-identified modules of technological knowledge both on the usage and supply side, moreover, makes it possible to grasp, especially at the aggregate level, the dynamics of increasing returns. The larger the number of agents which hold relevant portions of knowledge that are complementary, the larger is the output in terms of technological knowledge and eventually the wealth a system can generate. Externalities are directly the engine of increasing returns. In turn such increasing returns can be circumscribed within the boundaries of the knowledge modules.

This approach paves the way for a radical shift in the debate about intellectual property rights. A new chapter in the economics of intellectual property rights emerges. Intellectual property rights are now viewed as necessary signalling devices. Without intellectual property rights and specifically without patents, firms would rely upon secrecy in order to increase appropriability. Higher levels of secrecy in turn would make it more difficult to identify the relevant bits of knowledge and to activate the interactions which valorize technological externalities.

Patents are essential tools to signal the levels and the characteristics of the knowledge embodied in each organization. Patents are also more and more bargaining devices used by firms to improve their position when dealing with other firms engaged in complementary research activities. Patents are no longer regarded only as tools to increase appropriability, but also as devices to increase transparency in the knowledge markets and hence facilitate markets transactions. The signalling role of patents becomes relevant in this context as a device to reduce knowledge transaction costs. The build-up of reputation, by means of publications and scientific sociality, also plays an important role as a signalling device within the scientific community.

The informational role of patents as carriers of relevant information about the availability of new bits of knowledge in now more and more appreciated. The identification of each bit of complementary and useful knowledge as well as of the agents holding specific bits of knowledge and the assessment of their complementarity becomes an important function. This is expensive both in terms of search and opportunity costs: the costs of interacting with the wrong agents in terms of low opportunities. A specific form of knowledge transaction cost can be identified here. Patents are essential tools to signal the levels and the characteristics of the knowledge embodied in each organization.

The exclusivity of intellectual property rights is now questioned. The transition towards a system of interconnection rights, successfully tested in telecommunication networks, seems more and more necessary in the implementation of intellectual property rights. Perspective users of patents should find it easier to access the relevant property rights, provided that payments of royalties do take place. The shift towards the generalized use of compulsory licencing and of the liability rule might favor the systematic valorization of technological externalities (Dumont and Holmes 2002).

The relationship between external and internal knowledge becomes a key issue also in understanding the effects of external knowledge at the system level. It is immediately clear that substitutability cannot apply. Unconstrained complementarity, however, also appears inappropriate. The hypothesis of a constrained multiplicative relationship can be articulated. The ratio of internal to external knowledge seems relevant. Firms cannot generate new knowledge relying only on external or internal knowledge as an input. With an appropriate ratio of internal to external knowledge, instead, internal and external knowledge inputs enter into a constrained multiplicative production function. Both below and above the threshold of the appropriate combination of the complementary inputs the firm cannot achieve the maximum output (Audretsch *et al.* 1996; Veugelers and Cassiman 1999; Bonte 2003).

Because of the complementarity between internal and external knowledge, especially if it is specified in terms of a constrained multiplicative relationship, the aggregate outcome of both market transactions and interactions are unstable and sensitive to interactions and subjective decision-making. When both demand and supply schedules are influenced by externalities, multiple equilibria exist.

The amount of knowledge each firm can generate depends upon the amount of external knowledge available, that is upon the amount of knowledge that other firms, especially when involved in complementary research projects, have generated and cannot appropriate or are willing to exchange. The amount of external knowledge available at any point in time and in regional and technological space depends upon the amount of technological knowledge generated within the system and upon the conditions of technological communication within modules of complementary technological knowledge.

A very interesting case now emerges: in the markets for knowledge, both demand and supply externalities as well as joint-production apply and exert their effects. On the supply side, the amount of knowledge generated depends upon the innovative behaviours of the agents as well as on the general production levels of the economic system at each point in time and in the relevant past, because of the role of learning. On the demand side, as is quite clear, network externalities among knowledge users exert an ubiquitous role. The position and the slope of the demand schedule depend on the position and the slope of the supply schedule and vice versa. The latter in turn are influenced by the aggregate conditions of the economic system: learning rates depend upon the amount of output. Needless to say, however, aggregate output is influenced by the amount of technological knowledge generated in the system, via the total factor productivity effects.

At each point in time any solution can be found, but such solution has not the standard characteristics of stability and replicability. In the markets for technological knowledge each equilibrium point is erratic. Little shocks can push the system far away from the given values. No force will act to push the system back towards the levels experienced in the previous phase. At the heart of the market system, the production and the distribution of technological knowledge are characterized by multiple equilibria as well as micro-macro feedbacks and as such are sensitive to small and unintended shocks. Macroeconomic or monetary policies can have long-lasting consequences if and when they affect the joint-supply of experience and competence and hence the supply of technological knowledge. The strategic decision of firms to increase either the demand or the production of technological knowledge can also have long-lasting effects changing the parameters of the system. Entrepreneurial action hence may have direct consequences here at the economic system level, changing the equilibrium conditions. Both failure and success, however, can be the result, depending on the outcomes of the chain of reactions which may take place.

Economic systems may be trapped in a low-knowledge-generation regime, while others remain in high-knowledge-generation ones. Path-dependence, because of the role of learning and interdependence, deploys here its powerful effects. Small events can push the system to oscillate from one regime to the other with long-lasting consequences. In this context the issues of dynamic coordination among agents and institutions becomes most relevant in order to assess the general outcome of each single action.

Innovation, technological change and economic structure: a systemic view

About forty years after its inception, economics of innovation and new technologies is now sufficiently mature to fully acknowledge the contribution made by other social sciences in the definition of its basic categories and heuristic metaphors. At the same time, enriched by significant methodological extensions and unusually intense methodological and analytical discussions, the discipline can contribute, with its own specific tools and methodologies, to the study of the more general process of formation of new ideas in an advanced society.

The basic puzzle remains a problematic core for this interdisciplinary area of specialization. How innovations come to the market place, how novelty takes place in our understanding of the economic and technological interplay, how and why total factor productivity grows, how firms and economic agents at large generate and react to the introduction of novelty – these are still open questions.

The presentation of the shift from manna to product life cycles and trajectories and finally networks as heuristic metaphors for assessing the rate and direction of technological change can be aligned along two clear axes. The first runs from exogeneity towards endogeneity. The second runs from metaphors to concepts (Rosenberg 2000).

Along the first axis we see how the analysis of technological change as the result of an exogenous shock was consistent with an orthodox approach to neo-classical economics. All assumptions about endogeneity raised increasing problems about the dynamic laws of the economic system. The product life cycle approach first and then sequentially the technological trajectories and later on technological paths focused the analysis on the disequilibrium conditions in which innovation takes place and which innovation generates. The role of systemic analysis in understanding a dynamics of complementarity and interdependence – which cannot fully mediated by the price mechanism – characterize the network approach. From this viewpoint, economics of innovation seems to move away from the neo-classical textbook representation of the basic laws of motion of the economic system.

Along the second axis we see that each metaphor as a matter of fact has left the ground to more articulated economic concepts. Manna made it possible to explore the asymmetric effects of the exogenous introduction of new technologies. The product life cycle first, and the trajectory, paved the way to understanding the key role of cumulability, irreversibility and localized technological change. The network has led to a better understanding of the need for a systemic approach to analyzing the web of complementarities and interdependencies which relate each to other firms and technologies and cannot fully signaled by the price mechanism.

The basic assumptions about growth with constant returns to scale which made it possible to articulate the very notion of residual and hence total factor productivity growth remain at the core of the problem. The revival of increasing returns, properly blended with the results of the economics of innovation and new technologies, seems a promising direction for future research, one where the divide between the macroeconomics of growth and the microeconomics of innovation may be blended.

Cumulative technological change takes place, in out-of-equilibrium conditions, in an economic system where and when firms are not viewed as passive users of given technologies, only able to select the techniques more appropriate to a given set of relative prices, but as agents able to change and generate their own technologies. With these capabilities firms react to irreversibility traps and unexpected events both in product and factor markets by mobilizing appropriate levels of technological creativity.

A number of complementary conditions play a role. Firms are better able to change their technologies when, because of effective communication systems, local externalities can turn into collective knowledge; when high levels of investments can help the introduction of new technologies; when industrial dynamics in product and input markets can induce localized technological changes which in turn affect the competitive conditions of firms; when stochastic processes help the creative interaction of complementary new localized kinds of knowledge and new localized technologies to form new effective technological systems; when the dynamics of positive feedback can actually implement the sequences of learning along technological paths, as well as the interactions between innovation and diffusion. Such a set of dynamic and systemic conditions has strong stochastic features and is available in finite conditions: the process is unlikely to go on indefinitely to the exhaustion of the possible combinations. In these circumstances the generation of new technological knowledge and the introduction of new technologies can be viewed as the cause and the consequence of punctuated economic growth and increasing returns (Arrow 2000).

The successful accumulation of new technological knowledge, the eventual introduction of new and more productive technologies and their fast diffusion are in fact likely to take place in a self-propelling and spiraling process and at a faster pace within economic systems characterized by fast rates of growth where interaction, feedbacks and communication are swifter. The parallel emergence of the notion of co-evolution, elaborated in paleontology by Stephen Jay Gould (2002), here provides basic support to the notion of technological change which is more and more viewed as the end result of a process of dynamic interaction between the changing conditions of the structure of the system and the limited span of creative reaction of each agent, bounded by the effects of irreversibility as well as of its cognitive limitations, in its capability to explore the surrounding opportunities, and yet able to change its environment far beyond the standard price-quantity adjustments

Innovation is the result of out-of-equilibrium conditions and it is the cause of out-of-equilibrium conditions. A clear continuity ever since the biological grafts onto the trajectory and finally the systemic network approach confirms that innovation can only be understood in an analytical context which allows the integration of the analysis of firms and agents that are continually pushed away from potential equilibrium conditions and try and react to the unexpected conditions of both products and factors markets by means of the introduction of new products, new processes, new organizational modes, new markets.

Conclusions

The achievements of economics of innovation are now important enough to deserve a broader analytical framework to fully develop its heuristic capabilities. Economics of innovation has developed an important set of tools and arguments which help in understanding the mechanisms by means of which firms are able to generate new technological knowledge and to introduce and adopt new technological innovations.

The analysis of the determinants of technological change at the firm level makes an important progress as well as the understanding of the rate of introduction of technological innovations. The analysis of the direction of technological change instead receives less attention and the role of the structural context into which technological change is eventually introduced is less clear. At present, economics of innovation seems less and less able to reconcile the analysis of the rate of technological change with the analysis of its direction. Similarly, the analysis of the determinates is more clear at the firm level than at the system level.

The role of the structural characteristics of economic systems at large and specifically the role of the structure of relative prices, as determined by the endowment of basic inputs and the dynamics of industries and sectors, can provide economics of innovation with a context in which the understanding of the determinants and effects of technological innovations can be better appreciated and developed.

A long process has been taking place since the old days of knowledge as a public good. A better understanding has been elaborated of the dynamics of knowledge accumulation. Concern about appropriability conditions seems less and less relevant: because of the contextual character of knowledge, innovators are now thought to be able to appropriate relevant portions of the rents generated by the introduction of new knowledge. Demand and network externalities, however, play a much stronger role now. Transactions in the markets for knowledge do take place, together with technological interactions based upon barter and reciprocity. A variety of governance mechanisms has been designed and implemented, or simply better understood. The need for economic policy seems however stronger than ever. The governance of the markets for technological knowledge is not sufficient: a governance of knowledge commons needs to be implemented at the policy level.

The need for an out-of-equilibrium context of analysis, which is able to take into account the uneven dynamics of the full economic system, including the vertical and horizontal interactions among firms and between firms and households including factors markets, and to go beyond a theory of the innovative firm, seems evident. And this is the essence of the Schumpeterian legacy (Schumpeter 1928) which marks the economics of innovation.

The merging of the achievements of the "old" economics of technological change with the new achievements of the "young" economics of innovation can help such a process of recontextualization of the process by means of which new technological knowledge and new technologies are being generated and introduced. This approach can provide the basic elements in understanding the recursive causation of structural change and technological change and opening a new and fertile research ground to assess the process of economic growth and change. This broader context makes it possible to overcome the growing limitations of the economics of innovation which is more and more constrained in a competence theory of the firm and a decreasing heuristic capability to provide an overall analysis of the general process of growth and change at the system level.

A broader context of analysis such as this seems instead more and more necessary as many important achievements of the economics of innovation in terms of the complementarity between innovation and disequilibrium and the role of the institutional context in which firm operate and are able to innovate are being disregarded by the new growth theory. The new growth theory has been able to adopt and adapt much progress put forward by the economics of innovation and yet misses the core of the analysis: the path-dependent outcome of the interaction between the introduction of new technologies and the changes brought about in the economic system. Innovation is only a part of a broader process of interaction between the effects and the determinants of technological and structural change which takes place in a disequilibrium context.

3 The retrieval of the economics of technological change

Introduction

This chapter provides a synthetic summary and update of the most important results of the economics of technical change with special attention to the understanding of the effects of the introduction of new technologies. The economics of technical change in fact had well grasped the complex set of sequential changes which take place after the introduction of a new exogenous technology. When this analysis is extracted from its original static and single factors market framework and applied to a context characterized by heterogeneous and dynamic factors markets, important results can be obtained. The next section presents a selective retrieval of the most important achievements of the economics of technical change with respect to the analysis of the interactions between the direction of technological change of the system of relative prices. Pages 43–44 recall briefly the basic elements of the models of induced technological change. The conclusions summarize the main findings.

The rate and direction of technological change

In the economics of technical change tradition of analysis the introduction of technological change is portrayed as the substitution of a new technical space for the old one. A new map of isoquants is associated with a new technology. The introduction of a new map of isoquants has two consequences: first it may engender the substitution, with given factors prices, of production factors; second it makes it possible to increase the levels of output for given levels of inputs.

The analysis of the substitution of the new map of isoquants for the old one takes place in a single and static context: one where the heterogeneity of factors markets is not accounted for and the relative prices of production factors do not change. Moreover the case of possible overlappings between the old map and the new map is not considered. The analysis is concentrated upon the narrow region of techniques defined in the maps of isoquants defined by the previous equilibrium, as determined by the tangency between the isocost and the relevant isoquant.

In a first round much attention has been paid by economic analysis to the classification of technological change whether factor augmenting or neutral. According to the analysis of Hicks (1932) and Robinson (1937) technological change can be in fact neutral when it does not modify the ratio of the marginal productivity of capital and labor respectively. Conversely technological change will be biased, capital augmenting or labor saving, when the output elasticity of capital increases with respect to that of labor, and labor augmenting or capital saving, when instead the output elasticity of labor is augmented with respect to that of capital.

Other classifications of the direction of technological change have been elaborated by Harrod (1939) and Solow (1956 and 1957) and different definitions of neutrality have been provided. The former defines neutrality in terms of labor intensity of output; the latter in terms of capital intensity of output. In both cases technological change will be neutral when it does not change a key-input-intensity. These different definitions clearly overlap. A given technological change will be Harrod-neutral and capital-intensive according to Hicks; Solow-neutral and labor-intensive according to Hicks (Amendola 1976).

These definitions have been elaborated in the context of the neo-classical theory of distribution and assume the introduction of a single technology at each time and a single factor market in the system with a single wage and a single capital rental cost.[1] All the standard definitions of neutral and biased technological change assume a given and single market for inputs. No attention is given to understanding the possible variety of effects of technological change when it concerns a variety of regional factors markets and or their change in time.

After the path-breaking contribution of Abramovitz (1956), the classification of innovations has been enriched by a second dimension: the rate of technological change. Technological change can now be defined in terms of the increase of total factor productivity. An incremental technological change is defined by a small increase of total factor productivity. A significant increase of total factor productivity will define a radical technological change.

When we cross the different classifications of technological change and consider together the rate and direction of technological change, and attention is paid to the comparison of the full maps of isoquants, rather than to the region surrounding a given isocline, a very interesting case emerges: the case for technological variety. Technological variety emerges when a new technology is incremental, in Abramovitz's sense and biased (either labor-saving or capital-saving in Hick's sense). No technological variety takes place when technological change is: (a) either incremental or radical, but neutral; (b) biased, but radical.

Eckaus (1955/1958) is one the first economists to note that the analysis of the effects of technological change beyond the limitations of a single factors market context, yields important insights: "It is fairly common for observers to report finding modern capital-intensive equipment and techniques used in underdeveloped areas where relative factor prices would suggest the use of more labour-intensive techniques. I would now like to suggest that the use of the "modern" techniques is not necessarily irrational emulation but the result of real limitations in the technological choices available, and that this, in turn, is a major source of labour-employment problems in underdeveloped areas (1958: 354). As a matter of fact Eckaus does not develops all the implications of his insight and confines his analysis to the case of fixed coefficient techniques:

Suppose that, whatever the actual characteristics of the production function and degree of substitutability of factors, businessmen believe that they face a production function with constant coefficients, i.e., no factor substitution is possible. Indian businessmen, for example, may believe that the "American way" of producing is the best and only way and that this always involves high rations of capital to labour. Plant engineers accustomed to emulating "Western" technology may be not sensitive to the range of choice actually available in manufacturing processes and may impose technical constraints on managers in underdeveloped countries.

(Eckaus 1958: 353)

The points raised by Eckaus find a far broader context of application when the analysis of the interactions between the rate and the direction of technological change is developed in a global economic context, one where a variety of factors markets is considered. In this context, in fact, even if the degree of technical substitutability of a new technology is high, but the new technology is far superior to the pre-existing ones, local adopters in a country with endowments far different from those of innovators are forced to adopt and use factors proportions which yield less output than elsewhere.[2]

Following Momigliano (1975) these elements are relevant in a dynamic approach where heterogeneity of agents in space, and their evolution in time, and hence technological variety matter, one where a variety of factors and products markets can be assumed. This implies that different labor markets can be accounted for, according to regional, national and even industrial specificities. Hence different wage levels across countries, regions and industries can coexist. This is all the more relevant when a global economy is analyzed, one where firms located in different regions and countries interact in general markets. In global markets, capital rental costs can also differ widely, especially when we assume that the prices for capital goods can differ across regional markets. Moreover capital rental costs differ when barriers to entry and related extraprofits are at work. For different levels of barriers to entry and different equilibrium levels, within barriers to entry, in each specific markets, capital rental costs differ even for quasi-homogeneous interest rates (Momigliano 1975).

In such a context the conditions of use and adoption of new technologies, especially when the rate of technological advance, measured in terms of total factor productivity, is incremental and Hicks-non-neutral, can differ substantially across agents, depending on their location with respect to both factor markets such as intermediary inputs, capital markets and especially to labor markets, and to their product markets in terms of their scope of activity and industrial specialization (Antonelli 1995). Hence the need to consider a broad range of relative prices for both production factors and products and hence a variety of possible equilibria. When multiple equilibria exist the comparison between technologies requires that the full map of isoquants is confronted.

The definition of technological change as locally progressive and regressive and its appreciation with respect to the conditions for technological substitution can provide a workable framework of analysis able to generate set of hypotheses to try and understand a number of relevant dynamic processes. Technological variety takes place when a new technology brought forward by an incremental and biased innovation is introduced; it can be, in different factor market conditions, only locally progressive and regressive, rather than generally neutral and hence superior .

Rarely can the new technology be everywhere superior to the previous one. The sheer shape of any new isoquant, because of the bias, shows that it is superior (at the left of the equivalent one of the old technology) only for a limited set of techniques. In a single factors markets this is not relevant, but in global economy this has important effects. When a new technology is introduced, the new map of isoquants which accounts for the new technology presents systematic overlapping with the previous map. Graphically we see in fact that it is always possible that two identical levels of outputs can be obtained with two different isoquants, that is two different sets of alternative techniques. This takes place when a "small" shift effect – leading to a limited increase in total factor productivity – is accompanied by a bias, i.e. a change in the direction and the new technology is either labor or capital augmenting.

More specifically, with incremental technological changes which are not Hicks-neutral the overlapping of identical isoquants extracted from two maps of technology is not accidental, but, on the contrary, systematic. In these conditions the new technology, for any given level of output, and for any given isoquant, will be only locally more productive than the older one. As a matter of fact this new technology is not always better or superior to the previous one. It will be "better or superior" beyond the intersection point and "worst and inferior" before the intersection point. In other words such a technological change will be locally progressive as well as locally regressive.

This brings to the fore the need for a meta-map of isoquants where both (all) overlapping technologies are represented so as to take into proper consideration the full set of available techniques which agents consider when assessing the best production conditions, under the constraints of their specific and idiosyncratic market conditions.

In these circumstances we need to create an analytical space which can accommodate the behavior of agents which, at any point in time, can consider the choice between techniques which belong to different technologies. With over-lapping technologies agents face the need to confront the relative efficiency of techniques belonging to all (both) technologies at the same time.

Technologies can not be longer presented sequentially in historic time ordered along a well-defined hierarchical scale because technological variety and hence technological substitution takes place: technologies can be ranked only for given levels of factors prices. All changes of factors prices in time and in space can change the ranking of technologies.

These findings make it possible to appreciate a second and most important contribution to this line of analysis provided by the capital controversy. In the context of the theory of growth, Pasinetti has shown that there is a bijective relationship between distribution and production where the changes in the distribution of the revenue have a direct effect on the output (Pasinetti 1962 and 1981). The retrieval of the main findings of that debate can yield important, and still partly unexploited, results when applied within an industrial dynamics framework of analysis.

Important results can be obtained when a more dynamic microeconomic level of analysis is applied, one where technological change is analyzed in a complex context of analysis. This is where different factors markets exist and where technological change is the result of the endogenous efforts of a variety of players in a variety of industries and regions and where the vertical and horizontal relations among firms, industries, and regions within complex economic systems is taken into account. Within an industrial dynamics context it is clear that sensitivity of costs and productivity to factor intensity has many important implications. The level of output is influenced by the relative costs of all production factors. The dynamics of prices of intermediary inputs and capital goods provided by other industries, in other regions, has important effects on each system (Momigliano 1975).

According to Pasinetti (1962), all changes in the relative costs of production factors – as distinct from their absolute levels – have a direct positive effect on output. Such effects depend upon the ratio of the marginal productivity of that factor to the marginal productivity of the other factors. This is clear in a simple two-factors production function. Let us assume a highly capital-intensive production function in value added. The reduction in the relative cost of capital induces a substitution of capital to labor. Additional units of the more productive factor are added to the production process. Additional and more productive capital substitutes for less productive labor. Exactly the same process of course applies when the production function is labor intensive and a reduction in relative wages is accounted for.

In general the larger the productivity of the factor which is more widely used, the lower is the productivity of the factor which is less used and the greater are the effects of any changes in the relative levels of factor costs. Composition effects are the outcome of the sensitivity of output to the relative scale of each single factor, rather than to the scale of the bundle of production factors. They are positive when the relative scale of most productive factors is augmented and that of the least productive factors is reduced.

The composition effects of the changes in the relative scale of equilibrium use of each production factor can interact with the changes associated with growth and development. In a growth process many changes take place. Additional inputs can become available, and the relative costs of production factors can change. Also, and most important, structural changes in the organization and industrial dynamics in the upstream supply of production factors must be considered. In these conditions the composition effects can be very complex. Most important in this context is the introduction of new technologies with a significant bias in the output elasticity of production factors. Specifically the introduction of biased technological changes affects both the partial productivity of each production factor and the general efficiency of the production process. Structural changes in the vertical industrial organization can interact with the potential effects of technological change. The effects of biased technological changes, in terms of higher levels of total factor productivity, can be obscured by such changes in the "correct" combination of production factors.

The composition effects have also major consequences in generating significant international asymmetries. The performances of industries and firms within countries and regions, exposed to similar shocks can vary in very different ways, according to the consequences of the composition effects and ultimately to the characteristics of their production functions and the local structure of endowments.

The analysis provided by Wilfred Salter in the path-breaking *Productivity and Technical change* (1960) remains one of the highest contributions to the economics of technical change.

Salter focuses his analysis on the "interplay between technical advance and changing factors prices" (Salter 1960: 27). According to Salter the overall effects stemming from the introduction of a new technology are the result of "four components, three of which relate to the characteristics of technical advance, and the fourth to changing factors prices" (ibid.). The first effect identified by Salter is the neutral shift towards the origins of the production function. The second effect is the bias, "the movement of the production function may be greater towards one axis than another". The third effect is represented by the elasticity of substitution, that is the steepness of the curvature of the isoquants. Changing relative prices, finally, represent the fourth influence. According to Salter:

> The observed movements of best practice are the net result of these four influences – broadly represented by the rate of technical advance, bias towards uneven factors saving, the ease of substitution and relative factor prices – it is important to give them some precision in order to distinguish the nature of their separate influence and the interaction between them. But to do so is complicated for two reasons. First, the process of technical advance and factor substitution are so interwoven that we can only hope to distinguish their separate influences at a highly abstract level.
>
> (Salter 1960: 29)

With the contribution of Wilfred Salter the economics of technical change produces an unparalleled result. These achievements should now be retrieved into an analytical context where technological change cannot any longer be regarded as an exogenous event. The deep understanding of the fabric of effects of technical change can now be regarded as a necessary and complementary contribution to the general analysis of the incentives and constraints that the characteristics of economic systems impose as guiding mechanisms that induce and shape the direction of technological change and of the feedbacks that the introduction of new technologies engenders in the structure of system itself.

The models of induced technological change

Economics of technological change provides a superb line of enquiry into the analysis of the relationships between relative prices and technological change.

John Hicks, in his celebrated *Theory of Wages* (1932), provides the first basic insight into the theory of induced technological change:

> The real reason for the predominance of labor saving inventions is surely that . . . a change in the relative prices of the factors of production is itself a spur to innovations and inventions of a particular kind – directed at economizing the use of a factor which has become relatively expensive.
>
> (Quoted by Ruttan 2001: 102)

Along these lines two arguments have been elaborated. The first, often referred to as the growth theoretic model, elaborates the argument that with given factors prices, given research budgets and a frontier of possible technological changes, either labor-saving or capital-saving, firms have an incentive to introduce technological changes which make it possible to reduce the requirements of the production factor which takes the largest share of the revenue (Kennedy 1966). The model elaborated by Kennedy (1966) makes use of an application of the frontier of production possibilities where the possibilities of introducing either capital-saving or labor-saving technologies are directly confronted. The distribution of income between production factors – as shaped by the existing technology – defines the slope of the isorevenue. Standard maximization makes it possible to select the best direction of technological change. This analysis is clearly very abstract and essentially static in character. It is not clear why firms are actually induced to change their technology and to use resources in this activity.

A more fertile direction of analysis is paved by the contributions of Ahmad (1966) and Binswanger and Ruttan (1978). The analysis is now set at the microeconomic level and the flavor of the Hicksian suggestion is fully developed. The basic assumptions of this approach are the following: (1) an innovation possibility curve – stylized as a meta-isoquant – exists and firms have access to it; (2) firms are induced to change the technology when relative factors prices change; (3) the movements on the innovation possibility curve and hence the direction of the technological change are set by the kind of change in the relative prices. When the relative price of the production factor X increases, firms are induced to move along the innovation possibility curve and introduce an X-saving innovation.

The weaknesses of these models have been often stressed. The key elements of the economics of innovation are completely missing and the analysis of the actual availability of technological knowledge and technological innovations are not problematized.

Yet the inducement models do provide a clear guide in analyzing the determinants of the direction of technological change and the relations between the levels and changes of relative prices and the kind of new technologies being successfully applied in a given economic system, in terms of the relative bias. The incentive to introduce technologies which make a more intensive use of some production factors is closely related to the structure of relative prices and their dynamics.

Conclusions

The economics of technological change has been elaborated in a context of analysis heavily characterized by the attempt to provide a first dynamic flavor to the neoclassical tradition. The production function was the single tool used to set a stylized analysis of technologies. Economics of innovation dug much deeper into the origins and the dynamics of the processes that make it possible to generate new technological change and new technologies. Yet the retrieval of the economics of technological change seems useful and necessary, now, when the issues related to the direction of technological change emerge again as an important and somewhat forgotten area of research.

The economics of technical change has provided important tools to understand the effects of the introduction of technological change in terms of factor substitution and general efficiency levels. In this tradition of analysis, however, technology and relative prices can interact with each other in assessing output levels, average costs and the actual levels of total factor productivity. When a global and dynamic context is considered, one where there is a variety of factors markets and they change in time it is clear that the rate and direction of technological change can interact: a technology can be both locally progressive and regressive with respect to another. When technological change is both incremental and factor-augmenting, the scope for a trade-off between price efficiency and output efficiency emerges. New technologies are no longer always "superior". With low wages and incremental capital-using technological change, old labor-intensive technologies can be more efficient. In such conditions small changes in factor markets and small alterations in the ratio of wages to capital rental costs can have drastic effects on equilibrium levels inducing out-of-equilibrium conditions with significant discontinuities in productivity and employment levels, in the demand for capital and in the adoption of specific capital goods.

The interaction between relative factors prices and technological change also takes place with respect to the generation of new technologies. According to the induced technological change approach, the levels of relative prices and their changes have a causal effect on both the rate and the direction of technological change.

The economics of technical change deserves attention for the important insights it provides into the relations between technological change and its context of introduction. Much current analysis of the effects and determinants of the new wave of technological change seems to miss the necessary systemic understanding of the static and dynamic role of the structural characteristics of the economic system into which the new technologies are being introduced. More generally, too much attention has been paid to assessing the effects and the determinants of the rates of technological change, while too little analysis has been devoted to understanding the determinants and the effects of the direction of technological change. Even more obscured has remained the issue of the interactions between the rate and the direction of technological change in a dynamic a complex context, one where factors costs are allowed to change in time and in space.

The understanding of both the rate and the direction of technological change and the appreciation of the structural context of introduction are necessary to assess the present wave of new information and communication technologies. The recent wave of innovation and new information and communication technologies seems to be the result of a rapid sequence of complementary and interdependent technological changes. These changes lead to the introduction of a new technological system which can be defined as skilled-labor-and-capital-intensive and unskilled-labor-saving as well as highly service-intensive with a global scope of application. The global character of the information and communication technologies makes the analysis of the interactions between the structure of the economic system and the direction of technological change all the more relevant. Here is compelling evidence for the relevance of the variety of specific factors markets, the single competitive arena for products and the variety and complementarity of sequential introduction of technological innovations and adaptations, characterized by a close proximity in terms of their technical requirements and specifications in terms of output elasticities.

Part II

Innovation and structural change

4 Composition effects

The direction of technological change and the context of its introduction

Introduction

Much attention has been paid in the economics of innovation to the rate of technological change. Much less analysis has been focused upon the direction of the new technologies being introduced and to the structural characteristics of the economic systems into which the new technologies are being introduced. As a matter of fact the direction and the rate of technological change interact in many ways with the context of introduction and its evolution, and affect in depth the actual effects of technological change. The rest of this chapter elaborates the analysis of composition effects. Pages 50–59 consider the effects of the direction of technological change in heterogeneous factors markets. Pages 59–61 consider the effects of changes in relative factors prices on average production costs with a given technology. The conclusions summarize the results and pave the way to further analysis.

Composition effects

Composition effects have major implications for the analysis of technological change across different industries and countries because of the strong effects of relative factors prices on the actual "measured" total factor productivity growth of each country. The static and dynamic interactions between types of changes in technology and levels and changes of the relative price of production factors are relevant.

When technological change is biased, the context of its introduction plays a key role in assessing its effects in terms of total factor productivity growth. When a new technology is biased, in that it favors the more intensive use of a production factor, the effects of its adoption in terms of productivity growth will be stronger, the more abundant and hence less expensive the production factor. This dynamics has major effects, in terms of emerging asymmetries among firms in the global competitive arena.

When the scope of introduction of a new technology is global and the global economy is heterogeneous, it cannot be neutral everywhere. Only technological changes, characterized by a bias, consistent with the structure of local endowments, can reinforce technological variety in international markets where the relative prices of inputs differ because of local factor market differences. The global introduction of a new radical and biased technology, on the contrary, can reduce technological variety with negative consequences on the structure of comparative advantages and hence on the distribution of the gains from trade in the global economy

The introduction of a global and hence necessarily biased technological change has powerful effects in terms of new asymmetries among potential adopters. When a new technology is biased, the increase in efficiency takes place in only a limited number of techniques. In these conditions the new and the old technology are likely to intersect. Before intersection, in absolute terms, the new technology is superior to the old technology and vice versa after intersection. According to their relative factors prices, some countries will be able to benefit more than others from the introduction of the same technology. In fact, some countries would not be able to benefit at all: the old technology is still better than the new one.

Such asymmetric effects are reinforced and amplified by the dynamics of relative prices. When, with a given biased technology, relative factors prices, as distinct from absolute factors costs levels, change, output per unit of output and hence average costs also change. Specifically all reductions in the costs of the most productive factor have direct effects in terms of a reduction of the production costs and an increase of output per unit of input. Such changes in production costs do not have effects on total factor productivity measures, but in any event, do have powerful consequences in respect of the competitive advantage on global markets of rival firms based in heterogeneous factors markets.

Specifically we see that with a capital-intensive technology in place all reductions in the relative costs of capital, even if compensated for by an increase in wages, increase output levels. Similarly when a labor-intensive technology is in place a reduction in the relative cost of labor engenders an increase in output per unit of input, even if purchasing power is held constant. On the contrary, all increases in the relative costs of capital, with a capital intensive technology in place, lead to a reduction in output. This is also the case when an increase in the relative levels of wages takes place with a labor-intensive technology in place.

The analysis of the dynamic and synchronic interactions between factor endowments, relative factors prices and the rate and the direction of technological change are complex enough to deserve two distinct approaches. In the following section the effects of the direction of technological change upon heterogeneous factors markets are considered. The following section analyses the effects of changes in relative prices, holding constant the technology.

Relative factors prices, the direction of technological change and productivity growth

The analysis of composition effects has a strong and direct relevance in a synchronic context where a variety of factor markets across countries and regions is accounted for. A new and radical-capital saving technology will have stronger positive effects in a labor-abundant region with low wages. This explains why this technology will diffuse faster in such regions. The incremental labor-saving technology will have stronger positive effects and diffuse faster in a capital-abundant region with low relative capital rental costs.

A brief formal analysis can help to make the point clearer.[1] Formally we can see two different standard Cobb-Douglas production functions i and j with constant returns to scale. The latter has a larger total factor productivity level but lower output elasticity of labor:

(1) $Y_i = A_1 f(K^a L^{1-a})$
(2) $Y_j = A_2 f(K^b L^{1-b})$

the assumptions are

(3) $Y_i = Y_j$
(4) $b > a$
(5) $A_2 > A_1$

When assumptions (3), (4) and (5) hold, it can be shown that the isoquants of the technologies i and j overlap, so that Y_i and Y_j intersect for:

(6) $K / L = (A_1 / A_2)^{1/b-a}$

In fact:

(7) $Y_i / L = A_1 K^a L^{1-a} / L$
(8) $Y_j / L = A_2 K^b L^{1-b} / L$
(9) $Y_i / L = A_1 (K/L)^a$
(10) $Y_j / L = A_2 (K/L)^b$
(11) $A_2 / A_1 (K/L)^b / (K/L)^a = 1$
(12) $A_2 / A_1 (K/L)^{b-a} = 1$

When the two technologies i and j are in such a relation, it is clear that the bias effect interacts with the shift effect. A non-neutral technology superior in terms of shift effects can be inferior because of the bias effect. The context of introduction matters in terms of the local structure of endowments and hence relative prices.

Equation (12) suggests that the overlapping can only take place technological change is Hicks-non-neutral. It clearly makes a major difference, however, whether the overlapping takes place in marginal regions of the new meta-map of isoquants or, instead, affects the central regions. Central regions, with respect to the map of isoquants, are clearly those where realistic values of the ratio of wages to capital rental costs are represented. In central regions actual relevant choices are made and hence relevant actual economic behaviors are likely to take place: when the conditions stated in equation (12) apply in central regions of the meta-map of isoquants, technological change can be defined as contingent and hence both locally regressive and locally progressive.[2]

In such conditions the economic analysis of the effects of small differences in relative prices of production factors can yield important insights into the differentiated consequences of the introduction of the same technology across countries and regions. By the same token it is also true that the same small changes in relative prices can have major consequences in terms of production costs in two countries using two different technologies (see the following paragraph).

Equation (12) is important because it makes clear that the larger the shift, the more peripheral are the overlapping regions. This confirms that the more relevant the absolute increase of total factor productivity levels, the wider the scope of instantaneous adoption and hence the smaller the levels of technological variety. Only agents active in extreme factor markets could reverse the adoption of such technologies.

Equation (12) provides the basic discriminator to assess whether a new technology is general or contingent. A technology is contingent when the extent of the neutral shift is smaller and the regions where the old technology deserves rational adoption are larger. Hence the smaller the neutral shift, the larger is the scope for technological variety, for given levels of variance in factors markets conditions across regions and industries.

Figure 4.1 shows clearly that the relative distances AA' and BB' on the respective isoclines between the equivalent isoquants extracted from the two technologies are equal. Figure 4.2 on the other hand shows that the distance AA' is relatively larger than the distance BB': technological change is more effective in the upper region

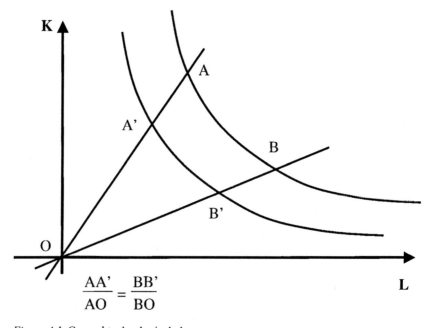

$$\frac{AA'}{AO} = \frac{BB'}{BO}$$

Figure 4.1 General technological change.

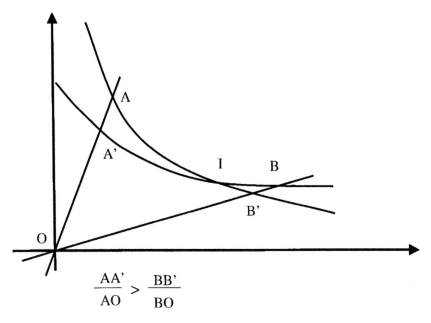

Figure 4.2 Contingent technological change.

than in the lower one. It is important to stress here that the distance BB" is actually negative while the distance AA' is positive. This is the basic difference with respect to the traditional analysis of factor intensity developed in the economics of technical change. The traditional definition of factor intensity has been elaborated with respect to a single system of relative prices, that is with respect to a given and single isocline. When the analysis takes into account the full metamap of new and old isoquants the coexistence of negative and positive distances emerge as a necessary condition. In Figure 4.2 the two equivalent isoquants eventually intersect.

In this context the specific characteristics of local factors markets play a major role. Because the isoquants of the different technologies overlap, it is always possible to find a specific isocost (see Figure 4.3) that is tangent to equivalent isoquants that belong to two different technologies. Formally in fact we see that:

(13) $W^*/R^* = dY_1/L^{1-a}/dY_1/dK^a = dY_2/L^{1-b}/dY_2/dK^b$

For all isocosts slopes that are smaller than W^*/R^* technology 1 will be superior to technology 2 and vice versa. A large set of techniques which belong to both technologies and are comprised between the two extreme values on the other hand will not be put into use.

It is now clear that a small change in relative factors prices can have major implications. Firms which rationally resisted the adoption of technology 2 will suddenly find it profitable, with major consequences in terms of performances and demand for production factors. Technology 1 is actually more productive not only before the intersection with the equivalent isoquant of technology 1 but before the tangency with the dividing isocost W^*/R^*.

It seems now clear that a biased technological change can affect different industries within different regions with different effects. The larger the variance in factors markets, the larger is the scope for technological variety. For a given new technology which makes intensive use of a production factor, industries and firms within industries located in regions where the most productive new factor is abundant are likely to be better off than industries and firms located in regions where the new most productive factor is scarce. Their total factor productivity is now ranked according to the relationship between the output elasticity of the production factors and the ratio of the relative costs of production factors.

Composition effect shapes total efficiency levels synchronically and also – and worse from an analytical viewpoint – diachronically. Technology 1 can rank better than technology 2 in a given country and for a given system of relative prices. When the latter change, however, the technological ranking may be reverted and the inferior technology 2 actually becomes better and vice versa. A strong case for technological variety emerges.

In a static context it is clear that for any given technology there would be a "best" system of relative prices and relative endowment of production factors. Specifically for a labor-intensive technology, a labor-abundant region would be the "best" factor market. Conversely for a given endowment and system of relative prices, there is a "best technology". For a capital-abundant region a capital intensive technology would clearly be the best one.

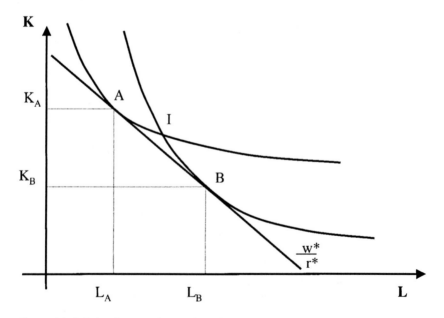

Figure 4.3 Relative factors prices and contingent technological change.

The analysis becomes much more complex when a dynamic context is taken into account: one where both technologies and relative prices change. A new understanding of the notion of technological change emerges from this line of enquiry. At this point it is important to distinguish between "contingent technological changes" and "general technological changes". The former consist in the introduction of a bias in the use of production factors, that is in changes in the shape of the technology without any shift, i.e. without changes in potential total factor productivity levels.

The combination of both movements, in the shape and in the position or level of the isoquant, that is the composition of both a shift and a bias, may lead to overlapping isoquants which belong respectively to new and old technologies. A meta-map of isoquants where both (all) technologies are represented becomes necessary.

The reference to a Cobb-Douglas production function here is useful. A new and actually more productive technology can be introduced without any actual increase in the parameter A which customarily measures total factor productivity levels. The technological change in this case is such that without any increase in the levels of the shift parameter an actual increase in total factor productivity levels may take place because of the new biased composition of the production function with respect to more and respectively less productive inputs.

The notion of contingent technological change differs from previous specifications of technological change. Technological change is neutral when it consists of a shift effect which leads to the traditional increase of total factor productivity levels with no effects in terms of the composition of the marginal productivity of the production factors. Contingent technological change instead affects only the composition and the ranking of production factors in terms of their output elasticity. The effects on total factor productivity are generated by the substitution of more productive inputs to less productive ones, with no shift in the production function.

A continuum can be identified between the two extremes of neutral/general and contingent technologies. At the one extreme we find neutral and radical new technologies that are consequently both general and global. Such technologies are characterized by such an important shift effects that they rank (almost) always and everywhere higher than previous technologies in terms of efficiency. Nevertheless they may be actually more productive in some systems than in others depending upon the relative costs of the most productive factors. The introduction of general technology with high levels of capital intensity in a capital-abundant country yields a larger increase in total factory productivity levels than in a labor-abundant country. It may still be adopted even in a labor-abundant country but it will exhibit lower levels of total factor productivity. The bias in the technology engenders a strong and long-lasting asymmetric effect.

A new general and neutral technology can be stylized as a production function where only the parameter A increases and the output elasticity of each production factor is not affected. With respect to the benchmark provided by equation (14) where the old technology is stylized, we see that with the new technology, stylized in equation (15) only A changes:

(14) $Y_{t1} = A_1 \, K^\alpha \, L^{1-\alpha}$
(15) $Y_{t2} = A_2 \, K^\alpha \, L^{1-\alpha}$

A general technological change consists of an actual increase of absolute total factor productivity levels. This increase is so strong and radical that even when and where the most productive factors are very expensive, actual total factor productivity levels increase with respect to (almost) any previous technique. In equilibrium models diffusion of these technologies should be instantaneous with no lags in adoption rates and no substantial variance in terms of penetration rates across agents in different regions and industries.

At the other extreme we find contingent technological changes. Contingent technological changes consist of innovations which can be stylized in a production function which fits better in each specific factor market. They consist of the single bias in the direction without any shift effect, i.e. without any change in the absolute total factor productivity levels. With respect to equation (14) now the contingent technology can be stylized so that the shift parameter A is not affected and only the output elasticity of production factors change:

(16) $Y_{t2} = A_1 \, K^\beta \, L^{1-\beta}$

Contingent technologies rank higher than previous technologies only in regions with similar local endowments: they are only locally superior.

In between the two extremes we can identify technological changes that consist of both a bias in the direction and a shift. Both make it possible to increase actual total factor productivity levels. Still with reference to the benchmark equation (14), now the new technology exhibits both a change in the shift parameter A and in the output elasticity of the production factors:

(17) $Y_{t2} = A_2 \, K^\beta \, L^{1-\beta}$

Such technologies rank higher than previous, older technologies only in a limited range of relative factors prices and hence only in a few factors markets. With a given system of relative prices such technologies are progressive. In other factors markets however they actually rank inferior to previous technologies. They are likely to be adopted by rational firms only in some circumstances.

It seems useful at this point to emphasize again an important result of this analysis: the ranking of new technologies depends upon the relative prices of production factors. Rarely can a technology be absolutely superior to any previous one. Hence technological reswitching can take place when the relative prices of production factors change. Technological reswitching is different from the classical technical reswitching. Here the ranking of technologies can be reversed by a change in the relative price of production factors. In the case of technical reswitching, the focus was rather on the ordering of techniques defined in terms of factors intensities. The actual levels of output which can be produced with a technology are influenced by the relative costs of production factors. The case of technological reswitching seems most relevant for technologies that are characterized by both a shift and a bias effect.

In the complex economic system where a variety of technologies complement the heterogeneity of endowments and local factors markets, the changes in both the technologies and the relative prices of production factors can affect the ranking of technologies. A technology which is neutral in the country of introduction in fact may reveal a strong bias in other adopting countries. From this viewpoint the distinction between shift and bias effects is blurred. The issue of technological reswitching may become relevant, especially in a global economy.

A distinction between potential and actual productivity growth can be introduced. Potential total factor productivity growth is obtained, for each given new technology, when the most productive input has the lowest cost. Actual total factor productivity growth is the one made possible in each region with the specific conditions of the local endowments. A full range of total factor productivity levels can be generated by the introduction of a single new technology in heterogeneous regions.

The actual ranking of technologies in terms of measured levels of total factor productivity depends upon the relative prices of production factors.[3] The effects of the introduction of a new technology, stylized by a new production function with a shift effect and different output elasticities for the two basic production factors, in two regions that differ in terms of factors costs, are larger, in terms of total factor productivity growth, the larger is the output elasticity of the cheaper production factor. Hence we must conclude that the standard general efficiency parameter, the output elasticity of labor and capital and relative factors prices matter in assessing the actual effects of technological change. The direction of technological change and the context of introduction are necessary components in a general assessment of technological change.

The received tradition of productivity accounting, based upon the path-breaking contributions of Abramovitz (1956) and Solow (1956), makes it possible to calculate a synthetic index of the changes in total factor productivity levels. With that methodology it is not possible to disentangle the composition effects, as determined by all changes in the relative prices of production factors and by the introduction of contingent technologies, from the shift effects.

Following Salter (1960) and Brown (1966) instead, simple calculations make it possible to decompose the standard residual and hence the total factor productivity level into two well-defined components: the effects of the introduction of general technologies and hence the shift effect, and the composition effects brought about by the introduction of new biased technologies which change the relative output elasticity of inputs.

The procedure is very simple and consists in calculating first the standard residual, based, as is well known, upon the calculation of a virtual output at time $t1$, based upon the new observed levels of inputs and the old output elasticities, and second, its comparison with the actual one. The difference is then attributed to the introduction of new technologies at large.

The complementary methodology, aimed at decomposing the bias and the shift effects, consists in calculating a new virtual output. The new virtual output is simply the product of the production function at time $t1$, with the new input levels and the new factors shares. The difference between the second virtual output and the actual one measures the shift effect. In turn the difference between the first virtual output and the second measures the composition effect.

Let us start again, with two simple production functions respectively at time $t1$ and $t2$. In the time interval a new technology has been introduced with both shift and bias effects and, moreover, relative prices have changed. Specifically we see that: the shift parameter has increased from $A_1 = 1$ to $A_2 = 2$. The output elasticity of capital at time $t1$ was $\alpha = 0.25$ and it is at time $t2$ $a = 0.75$.

(18) $Y_{t1} = A_1 \, K^\alpha L^\beta$ for $\alpha = 0.25$
(19) $Y_{t2} = A_2 \, K^a L^b$ for $a = 0.75$

The Abramovitz (A) residual is calculated as follows:

(20) $A\text{-RESIDUAL} = dY - (dY/dK) \, dK - (dY/dL) \, dL$

The shift residual (S) can now be calculated as the difference between the actual output and the estimated output expected when using the levels of inputs and the new output elasticities. Formally the calculation is as follows:

(21) $S\text{-RESIDUAL} = Y_{t2} - (K^{0.75} L^{0.25})$

In equation (21) the second term cannot include the effects of the changes in the shift parameter which are unknown. The output elasticities instead, with standard assumption about equilibrium conditions, can be derived from the share of production factors on income. The new levels of capital and labor are also drawn from the actual evidence.

The S-residual measures all the substitution effects, that is both the effects of changes in the relative prices of production factors and the effects of the introduction of biased technological changes which modify the relative productivity of inputs.

The difference between the A-residual and the S-residual can be termed C-residual, i.e. the composition residual which provides an indicator of the joint effects of the changes in the relative prices and in the output elasticities, and measures in a synthetic way the effects of the changes in the composition and relative efficiency of the production factors:

(22) $C\text{-RESIDUAL} = A\text{-RESIDUAL} - S\text{-RESIDUAL}$

It is important to note that the C-Residual may be negative as well as positive. A negative C-Residual takes place when a new general technology with a strong shift effect is introduced in a country although the factor intensities are at odds with the local conditions of factors markets. When the C-Residual is negative an important opportunity for the eventual introduction of dedicated contingent technologies emerges. The generation of new biased and localized technologies, that build around the new shift technology and make a more intensive use of the locally abundant inputs and hence save some locally scarce and costly inputs, may be very productive.

Relative factors prices and average production costs

Relative factors prices have a direct effect on production costs and output levels. When the technology in place is biased, production costs do reflect the structure of relative prices. Comparative advantage, among regions with heterogeneous factors costs and hence heterogeneous endowments are based upon the differences in production costs according to the differences in factors prices. In a dynamic context all changes in relative factors prices have direct effects on production costs. In a global open and competitive economy all reductions in relative factors costs, for the most productive input, have a direct effect on the levels of output and production costs. The extent of such effects is influenced by the bias of the technology in place. The stronger the bias, the more effective the dynamic effects of the relative factors costs upon the production costs. In turn production costs in the global markets have a direct bearing on markets shares and hence on opportunities for growth.

The formal analysis is here useful to clarify the point. Let us start with a simple Cobb-Douglas production function and the related cost equation:

(23) $X = K^a L^b$

(24) $C = wL + rK$

The dual transformation of the production function into a cost function, can be easily performed after taking into account respectively r and w, the unit costs of the two basic production factors, capital and labor. This leads to equation (25) where the long term dual average cost function has been derived from the production function:

(25) $C/X = w \left((r/w \, (b/1-b)) \right)^{1-b} + r \left((w/r) \, (1-b/b) \right)^b$

The differentiation of the dual cost function with respect to the ratio of the relative factor costs shows that output levels and hence average costs are sensitive to the ratio of factor costs. This effect is stronger the larger is the difference of the ratio of the output elasticities from unity. Formally we see that:

(26) $C/X = w(r/w \, b/1-b) \, (r/w \, b/1-b)^{-b} + r \, (r/w \, b/1-b)^{-b}$

(27) $C/X = (r/w \, b/1-b)^{-b} \, (r \, b/1-b + r) = (w/r \, 1-b/b)^b \, (r \, b/1-b + r)$

(28) $C/X = (w/r \, 1-b/b)^b \, (r/1-b)$

(29) $C/X = r \, (w/r)^b \, (1-b)^b - 1 \, / \, b^b$

(30) $d(C/X)/d(w/r) = b \, r(w/r)^{b-1} \, (1-b)^{b-1} \, / \, b^b$

From equation (30) it is clear that all changes in the ratio of the relative prices of production factors affect the average costs and that the effect is stronger the larger the difference from 1 of both the ratio of factors costs and the ratio of output elasticities. If wages equal capital rental costs there is no composition effect on average costs when, because of a biased new technology, the output elasticity of inputs changes. The same is true when the output elasticity of capital equals the output elasticity of labor. When the isoquant is perfectly symmetric and the slope of isocost equals unity, composition effects are nihil. Too often such an undergraduate textbook exposition is assumed as a legitimate generalization. As a matter of fact instead, and especially at the microeconomic level of analysis, technologies exhibit a significant bias and the differences in factors costs are relevant.

A simple numerical exercise makes the point clear. If the change in the relative prices is perfectly compensated so that the product of r and w is kept constant, average costs (AC) do not vary only when $a = 0.5$; AC vary instead in all the other cases. For $a = 0.3, r = 0.1, w = 10$, AC $= 0.73332$ and fetch the value AC $= 1.76601$ for $r = 0.9$ and $w = 1.11$. For $a = 0.5, r = 0.1, w = 10$, AC $= 2.0$ and stay at this level for all relative factor costs including $r = 0.9$ and $w = 1.11$. For $b = 0.7, r = 0.1,$ $w = 10$, AC $= 4.62695$ and reach the value AC $= 1.92131$ for $r = 0.9$ and $w = 1.11$. For $b = 0.9, r = 0.1, w = 10$, AC $= 8.73337$ and fetch the value AC $= 1.50587$ for $r = 0.9$ and $w = 1.11$.

The results of equation (30) are shown in Figure 4.4 where it is clear that for a capital-intensive technology the compensated change of relative factors prices with the proportionate increase of wages and decline of capital rental costs makes it possible to reach a tangency solution on isoquant 2 from the previous isoquant 1 without any actual increase in the purchasing power.

In assessing the actual efficiency, defined in terms of average costs, equation (30) has two important analytical implications. First, relative factor costs interact with absolute factor costs in assessing production costs. Second, relative factor

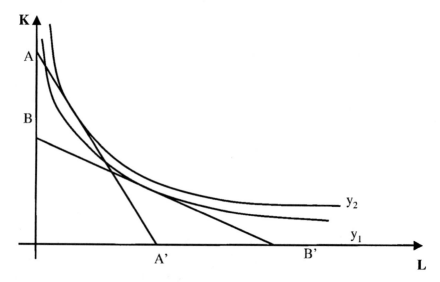

Figure 4.4 Relative factors prices and output.

costs bear effects on the general efficiency of agents using a given technology, when this is measured in terms of average costs.

From equations (29) and (30) it is clear that average costs can be low even if absolute factor costs are high, respectively lower than those possible with higher absolute costs, provided that the combination of the technology and the characteristics of the local endowments of production factors are such that the most productive input is relatively cheap.

A numerical example helps in grasping this relevant point. Let us assume the extreme case of a highly labor-intensive technology, say software, with $b = 0.9$. Average costs (AC) fetch the minimum 0.21952 for $r = 10$ and $w = 0.1$. With $r = 0.9$ and $w = 0.2$, that is far higher absolute costs, AC = 0.3218. For $r = 0.10$ and $w = 1$, AC = 1.742. With $r = 10$ and $w = 5$, AC = 7.4174. In this later case absolute costs are far above the benchmark and yet are still much lower than the extreme case of a factor market where $r = 0.1$ and $w = 10$, where AC = 8.73.

Absolute factors costs are compensated for by relative factors prices. Because relative prices compensate for the absolute level of factors costs they become a source of basic pecuniary externalities for firms. Competition among agents based in different factor markets is strongly affected by the relative prices.

The reduction of production costs engendered by the reduction of the relative prices, as distinct from the absolute levels of factors costs, is larger, the larger the range of output elasticities is. These effects in terms of production costs are important in a global contexts. Firms with lower production costs are more competitive and hence can acquire larger markets shares. This in turn provides opportunities for growth.

From a methodological viewpoint it clear that a case for total factor productivity growth cannot be made. An increase in total factor productivity cannot be statistically observed. Output per unit of input, however, increases, even if the technology has not been changed. Nevertheless, the firm, industry or region where the change in the relative prices has taken place is actually more efficient than before. The growth of nations and regions depends, also, upon the changes in the relative prices, for any given technology.

Conclusions

The analysis of the interactions between technological change and the structural characteristics of the economic system has made it possible to introduce an important distinction between general technological change consisting in the general shift of all the possible techniques, defined in terms of factors intensities, and contingent technologies which consist in a localized change of the mix of relative efficiency of production factors. Contingent technological changes engender a partial shift, while the shift brought about by general technological change concerns the whole range of possible techniques. This distinction can be better appreciated when the achievements of economics of innovation in understanding the determinants and the effects of the generation, introduction and diffusion of new technologies are considered in a single integrated analytical framework.

When a new biased technology is introduced in a heterogeneous economic system with a variety of local factors markets, the effects in terms of total factor productivity growth are influenced by the local structure of relative prices. The ranking of technologies is conditional to the relative prices.

When relative prices change and the technology in place is such that the output elasticity of each production factor is not the same, production costs and output levels also change. The reduction in the relative price of the most abundant factor has effects that are stronger, the stronger the difference in output elasticity is, with respect to all the other inputs. In the global economy the actual changes in the general efficiency of agents, in terms of average production costs, depend on both the increase in total factor productivity, brought about by new technologies in terms of bias and shift effects, and upon the changes in production costs brought about by the changes in the structure of the relative prices.

In a dynamic and global context, one where both relative prices and technologies can change and factors markets are heterogeneous, the general efficiency of each firm is influenced both by the changes in the technology and by the changes in the relative prices. The latter in turn is stronger the more biased the technological change is and has been. We can term these effects composition effects.

The direction of technological change and the context of introduction matter more than is currently appreciated, especially in a global economy, where agents based in heterogeneous factors markets compete in quasi-homogeneous products markets.

Two important notions can now be formed. First, the distinction between potential and actual total factor productivity growth. Potential total factor productivity growth is obtained, in a non-neutral production function, when the most productive input is cheapest. Second, the distinction between general efficiency and total factor productivity growth. Production costs, for a given technology, are influenced by the levels of relative inputs costs. The general efficiency of production can increase not only because of the introduction of a new technology, but also by means of a reduction of the relative price for the most productive input. The effects of given relative factors prices on the range between potential and actual total factor productivity levels and the consequences of the changes in the relative inputs prices on production costs, for a given technology, can be termed composition effects.

The generation of either contingent or general technological changes cannot any longer regarded as an exogenous event which takes place as the result of an autonomous process with no economic inducements and incentives. On the contrary, the introduction of both contingent and general technological changes can be considered as the outcome of quite specific incentives and constraints exerted and shaped by the structure of the economic system. Here the tradition of analysis built into the economics of innovation plays a key role in providing the necessary tools.

The identification of such structural incentives is a first step towards the full understanding and mapping of the path-dependent characteristics of the evolution of the system. Path-dependence is the result of the dynamic interdependence between the effects of the structure and its changes upon the rate and direction of technological change and the effects of the rate and direction of technological change upon the structure of the economic system.

5 New technologies and structural change
Constraints and inducements to innovation

Introduction

The analysis of the interaction between composition effects and technological change and the notions of general and contingent technological change have many important dynamic implications both for the economics of innovation and the economics of structural change.

A divide is consolidated in economics between the notion of technological change and the notion of innovation. The former is used to define the introduction of more productive techniques, with a given system of relative prices. The latter is frequently used to define all possible changes in the production and organization of the firm without any clear reference to their characterization in terms of factor intensity and the effects in terms of total factor productivity. As an important result of our analysis it seems now possible to reconcile these two strands of analysis. As a matter of fact, for a given system of relative prices, all changes to the spectrum of techniques in use have actual effects in terms of average costs and hence in the relationship between inputs and outputs. The distinction itself between techniques and technologies is blurred.

This chapter provides a systematic analysis of the inducement mechanisms that lead firms to introduce new technologies, in a context where both the rate and the direction of technological change are considered. The next section presents a broader definition of the innovative firm which takes into account the role of relative factors costs in assessing the actual performances of new technologies. Pages 70–79 explore the determinants of the inducement mechanisms which lead firms to the introduction of either general or contingent technological changes. The conclusions summarize the main findings.

The localized generation of general and contingent technologies

Innovation consists in the introduction of techniques that make it possible to produce a given output with a new mix of production factors even outside the pre-existing isoquants so as to affect directly the performance of the firm, under the constraint of the absolute and relative price of production factors. Innovation consists in the capability to move in the space of techniques, beyond the specific shape of the boundaries of equivalence defined at each point in time by the maps of isoquants.

A firm is innovative and successful when and if it is able to appreciate the bijective relationship between the constraints of the technology in place and the constraints imposed by the local systems of relative prices. From a dynamic viewpoint the innovative firm is successful when it is able to master the coevolution of both the relative prices and the technology. The understanding of this relationship makes it possible to consider a broader range of innovations including both those generated by the application of new scientific discoveries and those consisting in the manipulation of the technology so as to make it better and more consistent with the structural characteristics of the economic system.

Technology and location interact in many ways. For each given technology and a variety of possible locations in different economic systems with different relative prices, there is always a best solution and consequently a ranking of locations. The best location clearly provides the most abundant supply of the most productive factor. Conversely it is also clear that for each location, and hence each system of relative prices, there is always a better technology. The ranking of technologies depends upon the output elasticity of the locally most abundant factor.

With respect to the theory of the firm this is most important because it stresses the central role of a variety of specific competencies. In order to achieve high levels of performance firms need to know not only how, but also where and when. The direction of innovation efforts is clearly influenced by the specific endowment of the economic system where each firm is embedded.

At the same time this approach stresses the limitations of the so-called competence-based theory of the firm. Too much emphasis is put on the entre-preneurial capability to innovate of single firms and to little attention is paid to the structural determinants of the successful introduction of innovations. A broader set of factors needs too be taken into account by the theory of the firm and specifically the role of relative factors prices and of location in economic space.

It is also important to note that a trade-off may consolidate between technological change and relocalization. Firms may always achieve higher levels of actual total factor productivity by changing the location of their production facilities to sites which provide a larger supply and hence lower relative prices of the most productive factor of a given technology. The choice of relocation may substitute for the introduction of new more productive technologies.

Globalization is both the result of institutional changes in the international political arena and of the increasing drive towards internationalization of companies via increased flows of export of their products, increased flows of imports of components and other intermediary inputs, and multinational growth, by means of foreign direct investments in regions which can make a better use of well-selected technologies.

The growth of multinational companies can now be interpreted as the result of the search for competitive advantage by firms which try and master both the technology and the relative factors prices. Multinational companies in fact provide the best example of agents able to manage the coevolution of both the structural characteristics of each economic system and the direction of technological change. The multinational global corporation in fact is a portfolio of technologies and countries where each location should provide the best match between the technology, in terms of relative productivity of each production factor, and the local relative prices.

The understanding of the range in total factor productivity levels engendered by the introduction of a single non-neutral technology in heterogeneous regions and of the effects of relative prices on average production costs, provides the economics of innovation with a broader perspective. The actual performances of the innovations introduced by each firm are strongly influenced by the specific characteristics of the economic system in which each firm is embedded. Too much emphasis has been put by economics of innovation on the firm as the single relevant unit of analysis. More attention should be paid to the role of the economic structure with special attention to the markets for both basic inputs and intermediary production factors and hence to the industrial architecture of each system, in order to grasp the characteristics and the effects of the interplay of the dynamics of technological and economic change.

The main results of the economics of innovation in our understanding of the localized inducement mechanisms that lead to the generation, introduction and adoption of innovation contribute the analysis of the generation of either general or contingent technological change.

Elaborating upon the notions of bounded rationality, local search and localized technological change, innovation is viewed as the result of a local search induced by the divergence between expectations and facts. Firms are myopic agents affected by bounded rationality and as such are unable to anticipate correctly all the possible states of the world. Myopic firms are not able to calculate rationally all the costs and benefits of the introduction of innovations, moreover they resist the introduction of all changes which would increase the burdens and the costly limitations of bounded rationality. Myopic agents, however, may be induced to innovate and introduce technological changes when the current state of affairs seems inappropriate and unexpected events take place.[1] Here even myopic firms are aware of the costs of not changing their productive and commercial set-up. The costs of non-changing are then confronted with the costs of the introduction of new technologies.

The introduction of technological changes in fact is not free and it is the result of intentional conduct to a large extent. Each firm however cannot be analyzed in isolation, as far as the generation of new technological knowledge and the introduction of new technologies is considered. The characteristics of the collective networks of innovators and the structure of interactive learning in which each firm is embedded play here a major role.[2]

Innovation and the introduction of new technologies are the result of reactive and sequential decision-making activated by disequilibrium in both product and factor markets. Changes in the relative and absolute prices of production factors, as well as in the demand conditions for their products, push firms away from expected equilibrium conditions. In order to face the mismatch between the actual production set, as defined by previous irreversible decisions, concerning both fixed capital and labor – based as they are upon necessary but myopic expectations – and the unexpected changes in products and factors markets, firms however can (also) change their technology and can no longer be regarded as just quantity or price adjusters.

The introduction of a new technology, however, requires the investment of dedicated resources to conduct research and development activities, to acquire external knowledge and take advantage of new technological opportunities, to accumulate and articulate the benefits of experience and to valorize the tacit knowledge acquired in repeated processes of learning by doing, learning by using, learning by interacting with consumers, learning by purchasing. In such a context all changes in market demand and in the relative price of production factors are coped with by firms only after some dedicated resources have been applied to search for a new more convenient routine. Consequently in this approach firms make sequential and yet myopic choices reacting to a sequence of "unexpected changes" in their business environment, brought about the introduction of innovation by other agents in both products and factors markets.[3]

The introduction of technological changes is viewed as the result of the innovative behavior of agents constrained by relevant irreversibility and switching costs which keep them within a limited technical region and prevent significant changes in inputs composition. Technological change is introduced locally by firms able to learn about the specific techniques in place and hence to improve them.

When irreversibility matters all changes in current business require some adjustment costs that are to be accounted for. In our approach firms are portrayed as agents whose behavior is constrained by the irreversible and clay character of a substantial portion of their material and immaterial capital and by their employment levels. Moreover the conduct of firms is affected by bounded rationality which implies strong limits on their capability to search and elaborate information about markets, techniques and technology. As a matter of fact competence constitutes the basic irreversible production factor. In turn competence is embodied both in the organization of the firm, in the stock of fixed capital, in the levels of human capital embodied in the existing employment relations, in relations with suppliers and customers and in the communication channels in place with the markets and within the company itself (Antonelli 1999 and 2001a).

Myopic firms are induced to cope with the dynamics of demand and factor prices by introducing technological innovations, and they make the adjustments to market fluctuations while retaining, as much as possible, the previous levels of inputs and hence change locally the technology, according to the relative costs of introducing innovations.

The identification of two quite distinct classes of technological changes with respect to their effects articulates the analysis on the generation side. Two quite distinct rationales can be articulated, drawing from the economics of innovation tradition of analysis, to understand the generation respectively of contingent and general technological changes.

Four baskets of factors matter here. The first draws on the distinction between top-down scientific opportunities and bottom-up technological opportunities.

Technological opportunities are mainly based upon the learning processes which draw on new scientific discoveries while scientific opportunities draw on new scientific advances. The second concerns the location of the sources of new knowledge whether they are internal to the economic system in which the firm is embedded or mainly external, in other regions and even other countries. In this context, the regime of intellectual property rights and the levels of international protection, as distinct from those of domestic protection, play an important role in that they shape the actual conditions of access to external technological knowledge. The third relevant axis is the distinction between learning processes, whether it consists more of learning by doing or learning by using capital and intermediary goods purchased from other industries often located abroad. The role of switching costs provides the fourth relevant basket of variables affecting the innovative conduct of the firms, with respect to the costs associated with all changes in the existing stocks of tangible and intangible capital and techniques, including the expertise of workers and the brand and reputation of the firm.

For a given set of incentives and constraints, technological change will be either general or contingent according to the specific values of the parameters for these factors. When top-down scientific opportunities emerge and the frontier of scientific knowledge is brought forwards by relevant scientific advances; when internal knowledge is more relevant than external knowledge, when learning by doing is more relevant than learning by using, and irreversibility is low as well as switching costs, firms are more likely to introduce general technological changes. On the the other hand, when technological opportunities matter more than scientific ones, when the major sources of technological knowledge are abroad, learning by using is more fertile than learning by doing, and irreversibility of production factors, both tangible and intangible, is higher, firms are more likely, for given innovation budgets, to introduce contingent technological changes rather than general ones.

General technological changes consist of a radical shift of the map of isoquants, such that all techniques are now more efficient. They can be thought to be the typical result of scientific breakthroughs and research activities in technological domains where agents are able to improve the productivity of a large array of techniques. A major and radical breakthrough leads to new general purpose technologies. General technological change is characterized by a significant shift effect and hence high levels of total factor productivity. The shift effects are such that the new technology is superior to most (all) technologies in place in terms of rates of growth of total factor productivity. General purpose technologies however are likely to reflect the specific and idiosyncratic factor endowment of innovators: they are only locally neutral. Hence locally abundant factors are likely also to be most productive. The introduction of general purpose technologies can be thought to be the outcome of the localized efforts of innovators aware of new scientific opportunities and able to induce a general shift in the map of isoquants. Nevertheless the new technology is likely to be locally neutral, that is to reflect their own original technical choices and hence factor intensity. Even general purpose technologies can engender significant spreads in terms of total factor productivity growth across countries and regions that are characterized by heterogeneous endowments.

Contingent technological change can be conceived of as the result of the incremental introduction of a myriad of small changes after the main shift effect has been generated. Contingent technologies are introduced by firms, facing unexpected changes in both products and factors markets, when the constraints of quasi-irreversibilities of fixed capital stocks are lower and hence the switching costs associated with all changes in factor intensities are less important. Markets for inputs are here more flexible, the capital intensity is lower and as such the role of inertia engendered by sunk costs: firms can change their combinations with some ease.[4] Next and most important, contingent technologies can be considered the result of incremental innovations mainly built upon learning by using procedures. Firms learn how to use new general technologies, especially when the latter are embodied in capital goods and intermediary inputs, and eventually are able to capitalize upon the new tacit knowledge. The access to external knowledge by means of user–producer interactions with advanced, but remote sellers, sellers of new capital goods and intermediary inputs can help adopting firms to invent around and improve the factor intensity of the new general technology.

The generation of contingent technologies can be considered as the result of a viable innovation strategy for firms which have limited resources to fund research budgets, rely more upon external knowledge, associated with processes of learning by using new inputs, operate in flexible factors markets and are able to improve and eventually adopt new technologies, mainly invented elsewhere.

Specifically, a sequence between general and contingent technological changes can be articulated. In this sequence, after the introduction of new general purpose and yet locally neutral technology in a leading country with idiosyncratic factors markets, diffusion takes place at fast rates across regions and industries because of the strong increase in total factor productivity levels which the adoption of the new technology makes possible. In so doing, however, the new general purpose technology is adopted also in countries and regions where the relative prices differ sharply from the original ones. New adopters and other followers will try and increase the benefits of the new technology introducing contingent technological changes that fit better with the local endowment of production factors. The introduction of contingent technologies builds mainly upon the preliminary introduction of radical and general ones.

Contingent technologies can be viewed as the result of learning processes associated with the use of new radical technologies. The overlapping of different generations of biased technologies generates localized bumps in the map of isoquants. The introduction of contingent technological changes can be thought of a single step into a dynamic process of adjusting and adapting the bias of a new general purpose technology which takes place in a variety of specific factors markets, according to the local relative endowments. Eventually a new well-shaped general production function, with strong symmetric properties, might emerge in the global market, as the result of a sequential introduction of contingent technologies.

The analyses elaborated in the product life cycle context and eventually general-ized by the lines of enquiry conducted within the framework of the localized technological change can find an important use here. The sequence between general and contingent technologies – as defined in terms of factor intensity – in fact may take place with specific features, where the distinctions between the early introduction of major innovations followed by a swarm of minor incremental ones, and the sequence between product and process innovation, can be successfully applied.

A model of localized inducement of the rate and the direction of technological change

This analysis makes it possible to consider the scope for a localized choice, at the firm level, between the introduction of a new locally neutral technology which only consists in a shift effect and a new technology which mainly consists in a bias.

Bounded rationality limits the capability of agents to elaborate correct expectations as to all the possible outcomes of their decisions. Firms need to make irreversible decisions and yet are not able to anticipate correctly all the possible consequences of their decisions in the long term. Bounded rationality leads to a myopic behavior, but does not prevent agents from being able to choose among alternatives, even if not all the possible consequences are clear at the onset.

Firms which are active in factors markets radically different from those of original introduction of a new locally neutral technology can take advantage of contingent technological strategies and direct the funds available for intentional learning and research activities towards the introduction of new technologies which build upon the shift already introduced and are mainly directed towards a change in the relative composition of the productive inputs.

At the other extreme firms which already operate in the proximity of the technological frontier with production functions which already valorize the local endowments and exhibit high levels of output elasticity for locally abundant production factors have no other chance but to elaborate technological strategies finalized to the introduction of actual shifts in the map of isoquants. Research activities directed towards the introduction of general technologies are a necessary outcome of such conditions.

Firms based in intermediate countries face the real opportunity to choose between a more-contingent and a more-general technological change. It is clear that the introduction of new general purpose technologies which exhibit the specific mix of output elasticities most convenient with local factors endowments is more profitable than the introduction of contingent technologies which improve the local efficiency of a new general purpose technology introduced elsewhere. The relative costs of the introduction of a radical shift-technology with respect to a bias-technology becomes a crucial factor affecting the choice of firms in intermediate countries.

The access conditions to scientific knowledge, both codified and tacit, play a major role here. When and if the academic and scientific infrastructure is in place and appropriate incentives are at work, and if technological communication between the research centers and the business community is also effective as well as the general institutional conditions for the production and use of new knowledge, especially in terms of intellectual property rights, and large scientific opportunities are available, firms may be better able to direct their research strategies towards the introduction of more general technologies. Similarly, the availability of technological districts and local clusters of firms specializing in complementary research and innovation activities may help such choices.

Important technological opportunities offered by the introduction of new general technologies, and yet biased, at least for local adopters, instead provide important incentives to direct research strategies towards the introduction of more contingent technologies. The conditions of access to external knowledge possessed by the providers of the new technology is very important here, as are all user–producer interactions which make possible the communication of tacit knowledge. High levels of protection of intellectual property rights in the global economy can prevent the necessary adaptation of new general technological knowledge and delay the introduction of contingent technologies in following countries. All incentives to swifter trade in technological know-how, however, building upon strong protection of intellectual property rights, may reduce such risks.

The third relevant parameter is provided by the specific conditions of the factors markets. In regions and industries where factors prices are very close, so that the ratio of relative prices is in the proximity of unity, so as the slope of the isocost and the former technology could be stylized as a symmetric production function, the incentive to introduce contingent technologies is clearly very low. In these regions research strategies of firms are necessarily directed towards the introduction of technologies which do not change the factor intensity and mainly consist of a neutral shift. By contrast, regions where the supply of a specific input is abundant and its derived demand very low do provide a unique set of opportunities to direct research strategies towards the introduction of contingent technologies.

Similarly, in regions where the market prices of production factors are very elastic to all increase in their demand, firms are likely to direct innovation strategies towards the introduction of neutral technologies. This means that a research strategy mainly directed towards the introduction and adoption of contingent technologies can hold until firms are active in regions where the current factor intensity is significantly different from that of countries where shift technologies have been introduced. The difference in relative prices between countries is a prime inducement factor in the selection of innovation strategies.

The choice between the introduction of general and contingent technologies, once the firm has been induced to innovate by the new and unexpected conditions in its product and factor markets, can be nicely encapsulated by the analytical framework of nested frontiers of possible adjustments and innovations and two isorevenues.

Firms are induced to change the layout of their production process by the mismatch between the expected factors and products markets condition and the actual ones. The firms however have made irreversible decisions concerning both fixed and human capital and all changes in the levels of inputs, with respect to their plans, are expensive. Adjustments are but necessary: the out-of-equilibrium conditions generated by the mismatch between planned and actual conditions in the markets place generates losses and opportunity costs that cannot be sustained in the long run.

In this model all changes in the production layout and hence all movements in the existing map of isoquants, either on a given isoquant or from an isoquant to another – but still in the same map – engender switching costs.[5] Formally we provide the following definition:

(1) $SW = Z\,(dK/K, dL/L)$,

where dK/K and dL/L are defined as the changes in the levels of irreversible inputs which are necessary in order to cope with the new unexpected levels of demand and factors prices and SW stands for switching costs.[6]

The firm can either adjust to the factors and products markets conditions changing her position in the existing space of techniques, defined by the existing technology, or react with the introduction of an innovation which makes it possible to change the technology and hence the space of techniques.[7] The firm is now set to consider the fundamental trade-off between the costs of switching engendered by technical changes in the existing technical space and the costs of introducing technological changes which reshape the technical space.

The introduction of a new technology is the result of research and learning activities. The resources available to face unexpected changes in the products and factors markets can be both used in the generation of either general or contingent technologies. The investment of the resources available leads in turn to research, learning and communication activities which translate into varying levels of generation of either general or contingent technologies according to the relative ease of introduction of either kind of new technologies.

The firm in other words faces two nested frontiers of possible changes in the face of the mismatch between expected and real markets conditions. The first frontier of possible changes is the frontier of possible adjustments which make it possible to compare the results of resources invested in either technical changes or technological ones. The second frontier compares the kinds of technological change, whether contingent or general. The first isorevenue is defined by the absolute levels of the revenue generated by all adjustment activities consisting in both the amount of losses that are saved by the introduction of new techniques and the increase in output made possible by the introduction of the new technologies respectively. The second isorevenue compares the revenue generated by either general or contingent technological changes.

Standard optimization procedures make it possible to jointly identify both the correct amount of technological change with respect to the levels of switching technical change and the ratio of biased technological change with respect to shift technological change. Specifically it is a case of maximization for a given isorevenue level set by the amount of adjustment costs that are necessary to reduce the mismatch between expected and actual markets conditions.

Formally we see the following relations:

(2) $TC = a$ (research activities)
(3) $tc = b$ (switching activities)

(4) $GTC = c$ (general research activities)
(5) $CTC = d$ (contingent research activities)

where TC measures the amount of technological innovation necessary to change the technical space and "tc" measures the amount of technical change necessary to move in the existing technical space; GTC measures the amount of shift technological change, and CTC measures the amount of biased technological change that can be generated with a given amount of innovation-dedicated resources.[8]

Let us now assume that a frontier of possible adjustments can be considered, such that for a given amount of resources necessary to face the mismatch, firms can generate an amount of either technical change (tc) or technological change (TC). Nested to the frontier of possible adjustments we find a frontier of possible innovations that can be obtained with the introduction of either general technologies (GTC) or new contingent technologies (CTC).

Formally this means that:

(6) $tc = e(TC)$
(7) $GTC = f(CTC)$

In order for standard optimization procedures to be operationalized two isorevenue functions need to be set. The first, defined as the revenue of adjustments (RA) compares the revenue that adjustments by switching in the technical space yield (SW), to the revenue of innovation (RI). The second isorevenue confronts the revenue generated by the introduction of general technological changes to the revenues generated by the introduction of contingent technological changes. Formally we see:

(8) $RA = s\ SW + t\ RI$
(9) $RI = r\ GTC + z\ CTC$

where s and t measure the unit revenue of switching and the unit revenue of innovation; r and z measure respectively the unit revenue of the amount of general and contingent technological change generated with the given amount of resources available for innovation and induced by the new and unexpected conditions of the products and factors markets.

The system of equation can be solved with the standard tangency solutions so as to define both the mix of contingent and general technological change which in each specific context firms are advised to select and the amount of innovation with respect to switching the may want to prefer. The system of equilibrium conditions is in fact:

(10) $\begin{cases} e'(TC) = t/s \\ f'(CTC) = z/r \end{cases}$

$$\text{subject to}^9 \begin{cases} TC = GTC + CTC \\ RI = rGTC + zCTC \end{cases}$$

The alternatives of the adjustment process are stylized in Figure 5.1 where the intercepts on the axes of the frontier of possible adjustments shows respectively the levels of technical and technological change measured in terms of distance in the input space, and the isorevenue is set at the level defined by the amount of total adjustments costs the firm needs to fund in order to cope with the mismatch between expected and real market conditions. The analysis of the choice between innovation strategies directed towards the introduction of shift technologies and biased technologies respectively is expressed by Figure 5.2 where the intercept of the vertical axis exhibits the levels of innovation that a strategy directed towards the introduction of a shift technology can yield with given resources, available for research activities. On the horizontal axis the intercept shows the levels of innovation that a strategy directed towards the introduction of contingent technologies can yield. The slope of the isorevenue can measure the relative gross profitability of either research strategy. The search for the equilibrium conditions makes explicit the rationale of the choice for perspective innovators.

The cases of either only technical change or only technological change and in turn perfectly general technological change or purely contingent technological change seem extreme solutions. Much real world experience can be found in between such extremes. Firms are induced to innovate by the mismatch between actual and expected conditions of their production set, necessarily built upon irreversible decisions taken on the basis of myopic expectations which are not met by the

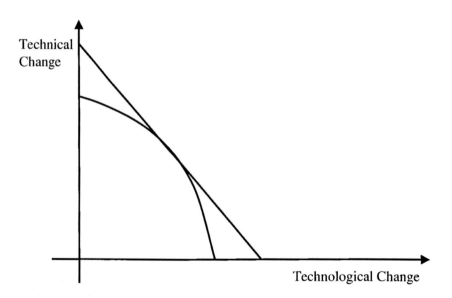

Figure 5.1 Technical change vs. technological change.

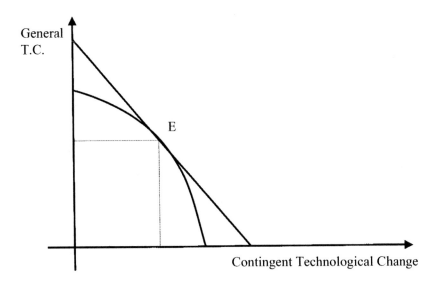

Figure 5.2 The choice of technological change.

disequilibrium conditions in product and factor markets. The direction of technological change is influenced by the relative profitability of introduction of general
technological change with respect to contingent technological innovations.

The correct direction of the new technologies being introduced can now be
considered as the result of two different but complementary processes. In an *ex
ante* perspective, myopic but creative firms select the kind of technological change
consisting of both a shift and a bias, which in the proper mix are most appropriate
to the specific conditions defined in the market-place both by the profitability of
the introduction of innovations and their relative cost of introduction, including
the levels of switching costs. In an *ex post* perspective, firms which by chance
introduced a technological change in the correct direction have higher chances of
surviving. Firms which introduce innovations with the wrong bias instead are likely
to be selected by the Darwinist selection mechanism activated in the products
market place by the rivalry among firms.

The analysis of the following chapters provides an in-depth assessment of
the different factors affecting the relative profitability and the relative costs
of introduction of either contingent or general technological change. A preliminary
analysis suggests that the profitability of introduction of contingent technologies is
positively affected by the barriers to entry and imitation that stem from composition
effects for competitors based in countries with different factors endowments.

In turn the introduction of general technologies can rely upon transient monopolistic extraprofits stemming from epidemic diffusion lags based upon information
asymmetries. Clearly the sharper the information asymmetry, the higher the
incentive to introduce general technologies. The long-term shape of the supply

schedule for production factors also matters here: the profitability of the introduction of contingent technological changes can be severely reduced by the steep supply of the most productive factors and hence the sharp increase of its relative costs because of the introduction of new technologies. Barriers to entry and to exit in upstream sectors may change the relative profitability of both the introduction and the adoption of new contingent technologies. In general it seems clear by now that industrial dynamics and markets structures play a major role in assessing the profitability of introduction of both technologies.

On the supply side the access conditions to external knowledge and the levels of switching costs are major determinants of the ease of introduction of either technology. The levels of irreversibility of fixed capital, both tangible and intangible, play a major role in this context because they affect directly the amount of resources that are necessary to manage the technical transition from one factor intensity to another, and as such, *ceteris paribus*, are not available for the generation of new contingent technologies. The generation of contingent technologies may be easier, from the supply side, but fewer resources are eventually available for their introduction.

The approach elaborated so far clearly belongs to the class of models of induced technological change. The inducement hypothesis works on the assumption that firms generate new technologies when factor costs change and, as in the post-keynesian tradition, when demand increases, at least with respect to their myopic view. Our approach differs from the standard inducement mechanism. Structural change here is at the origin of disequilibrium in both products and factors markets. New technologies have horizontal effects upon competitors and vertical effects on direct and indirect customers in downstream industries and direct and indirect suppliers in upstream industries, including labor and financial markets. Firms can cope with disequilibrium, in both factors and products markets, not only by adjusting quantities to prices and vice versa, but also, and mainly, by means of the generation, introduction and adoption of new technologies. Hence the primary inducement to introduce innovations is the disequilibrium in the market-place. This is the Schumpeterian legacy, much elaborated and enriched by the economics of innovation. However, the levels of relative prices and specifically composition effects exert a strong inducement on the direction of the new technologies being introduced. Relative factors prices induce the direction.[10]

The approach elaborated in this book also differs from the traditional inducement hypothesis, as articulated first by Hicks in 1932 with his path-breaking *The Theory of Wages*. Hicks paves the way for a tradition of analysis of the inducement hypothesis which builds upon the effects of the changes in the relative prices: no attention is paid to the levels of relative prices and to the composition effects. According to the basic hypothesis first introduced by Hicks and elaborated by Binswanger and Ruttan (1978) and recently updated by Ruttan (2001), firms introduce new technologies which save on the factor whose costs have increased. The inducement concerns both the direction and the intensity. An increase of wages in other words, in this class of models, is likely to induce the introduction of labor-saving new technologies. The steeper the increase of wages, the greater are the effects both in terms of labor-saving intensity and in terms of the amount of innovations being introduced.

In the approach elaborated in this book, instead, any increase in wages, as well as all changes in capital markets and in products markets, *per se* are likely to induce the generation of new technologies, because of the disequilibrium effects in the factors and products markets. Here the inducement to the rate is in place. The increase in wages, however, in a labor-abundant country with a large supply of labor and hence low wages should not induce the introduction of a labor-saving technology, but rather of a labor-intensive one, because of the powerful composition effects. The inducement to the direction is now different from that expected in traditional inducement models.

The identification of two quite distinct inducement mechanisms – the inducement to the introduction of innovations and the inducement to the direction of new technologies – seems relevant on three counts. First it provides a more articulated explanation of the substitution effect engendered by the introduction of new biased technologies. Second, it accommodates the post-keynesian inducement argument into a single integrated approach. Third, it remedies a basic inconsistency of the inducement hypothesis applied to factors markets where the prices of inputs differs sharply and the initial conditions of the production function are asymmetric. Let us analyze them in turn.

The distinction between inducement mechanisms seems able to provide a sensible answer to the well-known critique raised by Salter (1960) of the inducement hypothesis elaborated along the lines paved by Hicks (1932). Salter (1960) noted that firms should be equally eager to save on capital and labor irrespective of the recent increase in the unit costs of either factor. The basic aim of the firm is to reduce total costs. The approach elaborated here takes into account this argument. When relative prices change, firms are drawn into disequilibrium condition. Firms can either change their technology or their technique. Irreversibility and switching costs, however, induce firms to change their technology. The composition effects instead induce the direction of the new technologies. In order to increase output levels and reduce average costs firms will introduce and adopt the new technology which makes more intensive usage of the factor which is relatively cheaper. This direction-inducement mechanism is activated by the levels of relative prices rather than by their changes. All changes, instead, both in relative prices and demand, induce firms to innovate.

In the approach elaborated in this book, the inducement to the introduction and adoption of innovations, as distinct from the inducement of the direction of the new technologies, is not only activated by the changes in the relative factors prices, but also by all changes in the levels of demand. It is clear in fact that when demand levels differ from equilibrium ones, firms are induced to change their technology or their technique. For given switching costs they may want to change their technology. This is the classic kaldorian and generally post-keynesian demand pull effect, elaborated, in order to apply to the economics of innovation, by Schmookler (1966).

The distinction between inducement to innovate activated by disequilibrium conditions in factors and products markets, and the inducement to select a factor intensity for the new technology, seems able to reconcile different strands of the inducement hypothesis and provide a broader and coherent context into which they are complementary rather than alternative.

Strong assumptions about the full rationality and foresight of firms are not necessary. Myopic, but reactive and creative, firms can innovate in a variety of directions. Only the new technologies which make the best use of locally abundant production factors will survive in the product markets. Rivalry in products markets can be considered a reliable selection mechanism – a Schumpeterian Darwinism able to sort *ex post* the correct direction of technological change.

Finally, it is clear that when composition effects are taken into account the rudimental inducement hypothesis, according to which an increase in the unit cost of a factor (wages) should induce a specific factor saving (labor-saving) innovation, may be difficult to apply. The increase of wages in a labor-abundant country might induce the successful introduction of a labor-saving technology only if a strong shift effect also takes place. In such a country even if wages have just increased it still seems sensible to introduce labor-intensive technologies which take advantage of the low relative cost of labor. The basic hypothesis, as formulated by Hicks, can apply only in a symmetric and single system where both output elasticities and relative input costs are equal. The distinction between the inducement to innovate, stemming from all changes in demand levels and relative factors costs, and the inducement to direct the bias of the new technology, as dictated by the composition effects, seems able to save the inducement hypothesis from the poor assessment of the effects of asymmetric relative prices and output elasticities.

The framework elaborated so far provides a microeconomic understanding to appreciate the static and dynamic role of relative prices as determinants of the direction of technological change at the system level. The hypothesis that technology is not exogenous but is the result of the specific market conditions into which agents operate and reflects the historic process into which market interaction takes place, has been advanced repeatedly in the economic literature to explain the direction of technological change at the system level.

Habakkuk (1962) already articulated the hypothesis that American technology was different from the British as the result of the disparity of factor endowments in the two countries. The American economy was characterized by the substantial scarcity of unskilled labor and the relative abundance of natural resources and skilled labor. The British economy was instead characterized by the abundance of unskilled labor and the institutional and geographic scarcity of land and natural resources. According to Habakkuk, this disparity led not only to the obvious variety of factor intensities in the two countries, but also, and most importantly, to diverse paths of technological change. American technology was intrinsically biased in a labor-saving direction, while in British technology, was by capital-saving. David (1975) has further elaborated this frame of analysis suggesting that economic systems are better able to move along technological paths that push them to enhance their technology, following and deepening the original bias.

This argument, originally put forward by Habakkuk and David, has been the object of recent and systematic analyses according to which each system is able to introduce new technologies which are locally progressive and are localized in the range of techniques, defined in terms of factor intensity, that reflect the relative scarcity of production factors (Antonelli 1995, 1999a and 2001a). In this approach technology is endogenous and its direction is strongly path-dependent. According to this line of analysis technological efficiency is very much contingent upon its specific context of application. Each technology and the related bundle of techniques, defined in terms of factor intensity, is appropriate to a set of idiosyncratic market conditions.

Conclusions

Because of composition effects, the actual levels of measured total factor productivity of each technology depend upon the specific system of relative prices in each factors market. The direction of technological change in each regional system, characterized by a specific system of relative factors prices, can be affected by the composition effects in two ways.

First, the introduction of new technologies is induced by the disequilibrium conditions brought about in each economic system by the structural change which follows the introduction of previous technologies and in general by all changes in relative factors prices and expected demand levels.

Second, for a given inducement to introduce technological innovations firms in each region select the technology which fits better with the specific conditions of the factors markets. Relative factor prices become a selective mechanism which makes it possible to select technologies. Over time a region will make consistent choices and select technologies shaped by similar factor bias. Hence composition effects can be endogenized by perspective innovators which direct their tech-nological efforts towards the introduction of technologies which are specifically biased in such a way that they can make the best and more productive use of the production factors which are better available and hence have a lower costs in each specific region. On a general scale technological variety across regions emerges in both cases as the result of, respectively, the bias in the adoption and the bias in the generation of new technologies that are better suited to the specific markets for production factors in each region.

Such a bias in the direction of technological change can be seen as the result of an intentional *ex ante* decision of innovators well aware of the relative scarcity of production factors available in their own inputs markets. Innovative firms, for the given costs of an innovation, will find it more profitable to introduce new technologies which make a more intensive use of the locally most abundant factor. The bias in the direction of technological change can be also determined *ex post* by a selection process among innovators. Those who happened, by chance, to have introduced the technologies which are more intensive in the locally most abundant production factor would be sorted as the winners of the selection process. The replicator dynamics would force the "wrong" innovators out of the market and would favor the fast increase of the market shares of the "correct" innovators.

The direction of technological change in terms of the specific form of the bias sequentially introduced and adopted reflects the specific conditions of local factor markets. Well-defined technological paths emerge in each region in the long term as the result of the selection process in the general products markets. The more rigid and idiosyncratic the endowment of production factors and the system of relative prices, the more specific is likely to be the technological path of each region.

The model elaborated here provides a synthesis of the notions of internal and external path-dependence. Internal path-dependence takes place when the path along which the firm acts is determined by the irreversibility of its production factors and the opportunities for learning associated with the techniques in place. According to Paul David (1975) the choice of the new technology is influenced by the switching costs firms face when they try and change the levels of their inputs: firms are induced to follow a path of technological change by their internal characteristics. External path-dependence is determined by external conditions. Brian Arthur (1989) and Paul David (1985) have made the case for external path-dependence when new technologies are sorted by increasing returns to adoption at the system level. The model advanced here elaborates both upon internal and internal path-dependence. Internal path-dependence is appreciated because of the role of irreversibility and switching costs that are specific and internal to each firm. External path-dependence is contributed by the role of factors endowments and relative prices that induce the direction of technological change.

The exposure of each economic system to international competition, however, may change the direction of the technological path. After a new radical and general technology has been introduced in each country the search for appropriate technologies may lead to the introduction of new contingent technologies, that is the reshaping of the production function, without any actual increase in potential total factor productivity levels. In more successful cases the new technology can be general, that is, it can be both non-neutral and yet generally progressive.

In any event the introduction of new technologies is clearly the result of an out-of-equilibrium context which pushes the firm to the innovative choice, provided a number of key systemic conditions are available. This context provides a unique opportunity to blend the result of much economics of innovation with the economics of technological change in order to assess the rate of introduction of innovations, together with the technological characterizations of new products and new processes with an analytical framework which elaborates the role of factor intensities and output elasticities. The distinctive element of economics of innovation, the out-of-equilibrium context of analysis, is in fact the basic common thread and the unifying element.

6 Industrial dynamics and technological change

Introduction

Technological change has relevant effects on industrial dynamics and industrial dynamics has effects on the rate and direction of technological change. Such effects are not instantaneous and have a strong impact on the structure of the economic system considered as a complex web of industries that provide each other with intermediary inputs and capital goods.

A multisectoral analysis makes it possible to integrate the results of much economics of innovation beyond the limits of the analysis of the horizontal effects of the introduction of new technologies. The introduction of new technologies concerns not only competitors, but also suppliers and customers.

When the study of the effects of the introduction of new technologies is expanded, taking into account the changes in the relations downstream on users and upstream on suppliers, the frame of the analysis becomes much more complex. Effects on competition and effects on technological choice and eventually productivity are intertwined and sequential as they take place over time and with continual reciprocal adjustments.

The effects of technological change on industrial structure are both horizontal and vertical. Horizontal effects consist of the changes in the competitive advantages associated with the introduction of new superior technologies with respect to competitors. Vertical effects consist of the direct and indirect changes brought about by the introduction of new technologies on both customers and suppliers. Let us analyze them in turn.

The horizontal effects of technological change

The large body of evidence built by economics of innovation as to the horizontal effects of the introduction of a new technology, within an industry, provides the basic reference.[1] Technological change is a powerful factor of barriers to entry and important limitations to perfect competition. A huge literature, from Schumpeter, has explored how the introduction of new technologies yields transient monopolistic power. Innovators can retain a limited control of the specific knowledge upon which technological change relies. Eventually, however, innovations can be imitated and newcomers enter the market-place, bringing well-known positive welfare effects in terms of reduction of market prices, increase in demand and production.

When technological change is viewed as the result of the inducement engendered by changes in the market conditions, with respect to the myopic expectations of firms, it is immediately clear how the introduction of an innovation within an industry can lead to the introduction of further innovations by other firms in the same industry. Firms cannot elaborate rational expectations about the introduction of innovations by their competitors, as such all innovations affect the plans of existing firms and push them in out of equilibrium conditions. Technological change in turn becomes a viable solution to restore the profitability affected by the introduction of somebody's else innovation.

A choice between imitation and innovation however can emerge. The firm is induced to change its technology either by imitating the new technology already introduced or by the introduction of a new one. Barriers to imitation impede imitation. Transient barriers to imitation last as long as the information on the new products and processes is kept secret and does not leak out. Firms are likely to innovate, as opposed to imitate, the higher the barriers to imitation and the appropriability of the technological knowledge upon which the new technology is built.

This analysis, however, seems to apply to a closely circumscribed set of conditions. The new technology is locally neutral and all competitors, both innovators and prospective adopters, have access to production factors with the same relative prices. This is the result of a large amount of empirical evidence based upon the analysis of domestic competition within homogeneous industries (see Klepper 1996; Klepper and Miller 1995; Klepper and Simons 2000).

Quite a different context emerges when global markets are considered and the variance in factors markets is taken into account together with the characteristics of technological change. It seems clear that when contingent technologies are introduced in regional and industrial markets characterized by large differences among production factors, technological change can act as a powerful determinant of long-lasting barriers to entry.

Contingent technological change can be considered a major factor in raising absolute cost barriers to entry. Contingent innovators can retain their market power in the long term if they behave strategically and fix limit prices. Limit prices, in order to make it possible to gain extra profits in the long term without inducing the entry of newcomers, are set in the proximity of the production costs of competitors (Sylos Labini 1956/1962 and 1984; Momigliano 1975). In this context, the neoSchumpeterian tradition of analysis on the role of technological change as a factor of long-lasting market asymmetries in industrial dynamics receives new strength. Contingent technological change in the global economy may be the source of long-lasting competitive advantages based upon absolute costs. Only the introduction of rival innovations can contrast them. Competition can take place only by means of the direct rivalry in the types and kinds of technological changes being introduced.

Imitation of contingent technologies is impeded by the actual lack of viable incentives for adopters. Even a blue-print technology available on the shelf will not be adopted if it consists of a contingent technology appropriate to the factors markets of remote rivals. In such circumstances prospective adopters will never imitate and adopt the new technology, not only because of the transient effects of information asymmetries, secrets or intellectual property rights, but simply because the new technology is not more productive in their specific factors markets. Even if adoption is profitable, moreover, adopters would experience lower levels of total factor productivity growth than those innovators "in the right place" would take advantage of.

The differences in relative prices prevent adoption and provide contingent innovators with a long-lasting advantage in the products markets. It now seems clear that the larger the differences in relative factor costs among competitors, the larger is the incentive for innovators to introduce contingent technologies rather than general ones.

The distinction between probit diffusion and epidemic diffusion, as elaborated by the economics of innovation, is useful and effective here. General technologies do take time to be adopted. Information asymmetries play a major role in this context and delay adoption by less informed and less expert adopters. Eventually, however, information spread and the incentive to adopt becomes clear to everybody. Contingent technologies instead can be adopted only by prospective users who can actually take advantage of that specific production mix. Agents who have access to radically different factors markets may rationally decide to never adopt: this is the case for probit diffusion where a dichotomy emerges.

Clearly the diffusion of locally neutral general technologies, that is technologies which are characterized by both a shift and a bias – at least for adopters active in factors markets that differ from those of innovators – is likely to exhibit both epidemic and probit diffusion types. Epidemic diffusion based upon the contagion effects of the spread of information applies to all parties for whom at each point in time the new technology is actually profitable. A share of potential adopters however would not adopt even if fully informed, because of the bias effects. Changes in relative prices however may induce further waves of epidemic diffusion. A sequence of epidemic-diffusion processes articulated by changes in the threshold of probit-diffusion processes, activated by changes in relative prices, can be considered a suitable framework in which to understand the distribution of adoptions over time of such technologies.

When the specific conditions of competition in the markets place are considered a new set of incentives and constraints emerges as a bundle of powerful factors in assessing the content of the research strategies of innovators. The introduction of general technologies is a direct incentive for remote competitors towards adaptive adoption, as distinct from sheer imitative adoption. Building upon imitation and by means of learning by using, they must eventually introduce new contingent technologies which make it possible for them to compensate for the lower levels of general efficiency and reduce their competitive disadvantage.

When instead a contingent technology is introduced and moreover competitors are based in regions and industries with significant differences in relative prices, adoption cannot take place because the technology is not profitable. Diffusion does not take place and followers have little opportunity to learn how to use the new technology. A widening gap among leaders and followers can take place. Competitors must innovate in order to survive in the market place.

This analysis shows how relevant is the integration of the approaches elaborated in the context of the economics of innovation with a proper consideration of the structural conditions of the markets and the understanding of the characteristics of the technology with respect to both factor bias and total factor productivity. The conditions for exploiting the notions of probit and epidemic diffusion, as well as of the competition among firms based upon shifting barriers to entry and the introduction of new products and new processes, can be better understood when the effects of the differences in relative prices and hence factor markets at large is properly taken into account.

In this broader understanding, horizontal effects consist in the changes of the market conditions within each industry after the introduction of a new and superior technology. Within each industry, incumbents are now exposed to a decline of their market shares and sales. In this context the basic inducement to generate technological change, in order to face the new adversities, is in place. Such an inducement can lead to the adoption of existing available technologies or to the introduction of new technologies. The introduction of new technologies is necessary especially when the adoption of new technologies, available but introduced elsewhere, because of the composition effects, is not likely to reduce the new emerging costs asymmetries.

The vertical effects of technological change: the key industries

As a direct extension of the analysis of composition effects, it is clear that output and average costs are sensitive to the relative prices of both basic and intermediary inputs. The elasticity of output to the relative prices of intermediary inputs has important dynamic implications in the "filiere" approach. The overall performances of an economic system are strongly influenced by its composition in terms of key industries. Key industries provide the rest of the system with capital and intermediary goods which enter the production of downstream activities and contribute their own innovation activities, both indirectly, affecting the generation of new knowledge and directly for their effects on the levels of average costs (DeBresson and Townsend 1978). The relative prices of the new and highly productive intermediary inputs and capital goods have an important role in assessing production costs of downstream industries and hence of all the system. In turn, the introduction of technological changes in downstream industries has major effects on the derived demand for intermediary inputs and hence on the evolution of the industrial organization in upstream industries.

The analysis of the vertical effects of relative prices complements and strengthens the taxonomy elaborated by Keith Pavitt (1984) about the interindustrial flows of technological knowledge. The analysis of technological change provides an array of examples on these matters. The introduction of synthetic fibres in the textile industry had major effects on the demand for traditional fibres and related textile machinery both in spinning and weaving, causing major changes in the industrial structure of supplying industries and ultimately, further downstream in the fashion industry. The introduction of plastics had strong effects in an array of downstream industries by activating the substitution of steel, wood and leather in the automobile, furniture and shoes industries. The advent of mobile telephony has today similar effects on the demand for fixed telecommunications services. Data communication in turn reshaped the demand for computers and hence the organization of the computer industry itself with the decline and sharp redefinition of the market for large mainframes and the growth of the demand and the supply of distributed computing facilities including telecommunication services and personal computers.

The architecture of an economic system and the structural changes brought about by the introduction of new technologies have important effects on the overall performances of the system. Key industries are those which provide to the rest of the system intermediary inputs that are strategic for the successful adoption and incremental implementation of new technologies in a large array of downstream sectors. The adoption of new technologies in downstream industries is conditional on the provision from key industries of key complementary inputs that are strictly necessary for the use of the new technologies. The larger the number of industries which receive from the upstream vendor the strategic inputs, the more relevant is the supply of key industries. Advanced telecommunications and information services today provide a clear example of key industries.

The lack of key industries in the architecture of an economic system has many important and negative consequences. Import duties and transportation costs increase the relative prices of such strategic inputs. As such they reduce the levels of the actual profitability of new technologies in downstream industries and increase their production costs. The lack of key industries delays the adoption of innovations because it reduces the positive effects of the new technologies when complementary inputs have the "correct" relative prices. From a more dynamic viewpoint user–producer relations play an important role in directing and stimulating the introduction of incremental innovations which are better suited to fit the localized and idiosyncratic needs of specific categories of users. Once more the lack of a local supplier of highly productive intermediary and capital goods may delay the introduction of appropriate localized and incremental technologies.

The timely provision at competitive prices of new highly productive inputs affects the whole structure of relative and absolute prices in downstream industries with important effects on the cost conditions of both direct and indirect users. All delays in the adjustment to competitive prices for new production factors have complex effects downstream, especially on a competitive basis when different economic systems with different adjustment dynamics confront each other on the global markets.[2]

The pattern of evolution of production costs in downstream industries, after the introduction of a new capital good or an intermediary production factor which embodies a significant innovation, is likely to be deeply affected both by the horizontal effects of the new technology in upstream markets and by the vertical consequences of such horizontal effects.

The vertical effects of technological change and industrial dynamics: diffusion and adaptation

The introduction of innovations in upstream industries has relevant effects in downstream industries. The supply of intermediary inputs is changed by the new technologies. Myopic firms in downstream industries, unable to foresee the eventual effects of the introduction of innovations in upstream industries, may be induced to change their own technology, as a tool to face the new market conditions. This is the first and clear vertical effect, activated by the inducement mechanism: technological change in upstream industries is likely to induce the introduction of new technologies in downstream industries, because of the changes in the supply of intermediary products. The introduction of innovations in downstream industries, however, may have similar effects in upstream industries when the demand for intermediary products is considered. Now the introduction of innovations in downstream industries is likely to induce the introduction of technological innovations in upstream industries because of the changes in the demand for intermediary products.

Industrial dynamics within upstream and downstream industries can make this vertical inducement process more complex. Monopolistic markets and barriers to entry and to exit both in downstream and in upstream industries have major negative effects on the rest of the system. The monopolistic mark-up in fact has a direct bearing on the relative prices of highly productive intermediary production factors and capital goods with a general reduction of output and increase in average costs in downstream industries and hence in the rest of the system.

The analysis of the time path along which the markets for intermediary inputs adjust to the new conditions for their use, determined by the introduction of new technologies, has received little attention in the literature. The understanding of the important role of composition effects instead highlights the role of the filiere analysis. The time profile and the dynamics of the adjustments in the upstream markets, induced by the introduction of new technologies in downstream industries, and vice versa, that is the adjustments in downstream industries induced by changes in upstream ones, are very important to understand the time path of the effects of a new technology in the system.

First of all it is clear that the vertical effects of the introduction of new technologies consist in the direct change of the relative prices of the production factors in downstream production processes. Firms in downstream industries are exposed to new and unexpected conditions in both products and factors markets. The basic conditions of the inducement mechanism is put in place. Firms must adjust to the economic environment and can consider the opportunity to introduce in turn new technologies which fit better in the new system of relative prices. When and if new technologies are already available, the vertical effects of upstream innovations leads to a new assessment of the profitability of their adoption.

In this context it is clear that the analysis of diffusion finds new arguments. The determinants of diffusion, that is the causes of the non-instantaneous adoption of new technologies, can be better understood when the distinction between contingent and general technological change is appreciated and the differences in factors markets, including intermediary ones, across regions and industries and even among firms, within industries and regions, are taken into account. Only general technologies are such that their adoption is always convenient. And yet for given information and learning costs even the adoption of general technologies can be delayed: in some countries general technologies are more productive than others. Contingent technologies on the other hand may become profitable only when and if specific combinations of factors costs emerge.

The complementarity between adoption and adaptation becomes here a central issue. Rarely are technological changes so drastic and radical that the new technology can be adopted rationally in any circumstance. As a matter of fact new technologies are likely to reflect the specific factors endowments of innovating regions: they are locally neutral. Such diffusion is rapid only in regions and industries where the specific characteristics of the products and factors markets are close enough to reduce the asymmetries in terms of relative productivity of inputs. Passive adoption of technologies conceived elsewhere may be substantially delayed by such asymmetries. Instead diffusion may be rapid when contingent technological changes are introduced sequentially after a radical new technology has been introduced. Here prospective adopters are able to carry on substantial adaptation activities which make it possible to change the input specification of the new technology and yet to retain the advantages in terms of shift effects. Effective diffusion then is the result of both the adoption of a new superior technology and of adaptation efforts directed at changing its idiosyncratic factor bias.

Here again it is clear that much empirical evidence elaborated by economics of innovation about the sequence between radical innovations and minor incremental ones, and especially the analysis of diffusion as a process of introduction of a variety of products closer to the needs of prospective users, can be put into a more articulated structural context where the differences in terms of relative prices and factors markets play an important role.

The analysis of the effects of international industrial dynamics on the diffusion of new biased technologies can be developed with a small geometric apparatus. Let us assume a distribution of niches of potential adopters. Transportation costs may be considered the distinctive factor aligning the groups along a dotted line. The adoption of the new technology by each group can be considered a clear example of a probit diffusion process: adoption takes place when the profitability of the new technology is actually superior to that of the previous or to that of the new rival technology.

The introduction of a new non-neutral technology leads to some substitution among inputs: old inputs are substituted by new ones. Profitability of adoption by each group of potential users is directly affected by the industrial dynamics in the two upstream industries providing the relevant intermediary inputs. The introduction of the new technology in the core region where no transportation costs

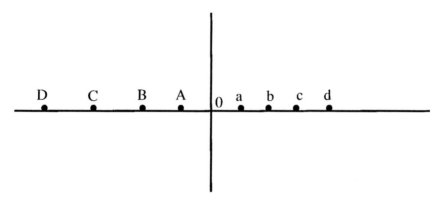

Figure 6.1 The distribution of the groups of potential adopters.

are to be accounted for has direct effects on the upstream industry which delivers the capital or intermediary goods. The supply of the new product is concentrated in few hands and incumbents enjoy substantial barriers to entry. The shift in the derived demand for the new product has the direct effect in keeping the market prices for the new good far above competitive levels, i.e. minimum average costs. Eventually imitation takes place and entry engenders the rightward shift of the supply curve so that prices decline towards the competitive levels.

The rates of imitation and entry and their corollary in terms of rates of reduction of market prices have a direct effect in terms of profitability of adoption down-stream. With the reduction of market prices new niches of adopters find the new technology profitable and adopt it.

By the same token the adoption of the new technology in the core region, with no transportation costs, has the effect of a reduction in the demand for the product

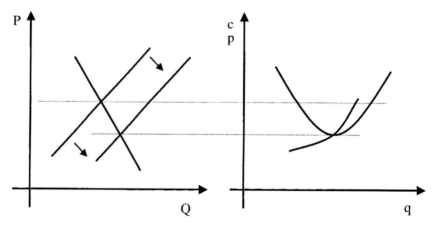

Figure 6.2 Industrial dynamics in the supply of new goods.

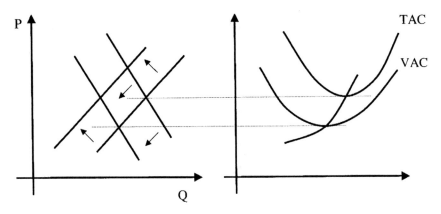

Figure 6.3 Industrial dynamics in the supply of old goods.

of the industry providing the "old" capital or intermediary inputs. The reduction in the levels of the demand induces a transient reduction in the market price. Such a reduction in the market price, however, is not yet such to induce incumbents to exit. The reduction in the price, below competitive levels, is such that incumbents are still able to pay variable costs. The market prices for the old inputs remain below competitive levels as long as incumbents are resilient.

The eventual exit of loss-making firms, however, restores competitive conditions and prices increase back to equilibrium levels. The duration of losses and resilience and the rates of exit have a direct bearing on the technological choices and the competitive advantage of downstream users. As long as the market prices for the old inputs remain below competitive levels some groups of potential adopters are discouraged from adopting the new technology and keep using the old one. The eventual exit of marginal producers will clearly have the direct effect of inducing new adoptions in downstream users.

When and if barriers to entry and exit respectively have perfectly symmetric effects in terms of transient market prices for intermediary and capital goods respectively above and below competitive prices, industrial dynamics affects relative factor costs but not absolute ones. The purchasing power of users is exactly the same. The compensation effects however do have effects because of the relative prices. When the effects are non-symmetric, however, absolute factors costs are also influenced by industrial dynamics and the vertical effects of the horizontal effects of technological change in upstream industries. The effects of the interplay between horizontal and vertical effects are all the stronger.

The diffusion of new both non-neutral and neutral technologies, defined as the sequential adoption by different groups of users, is clearly influenced by industrial dynamics in terms of rates of entry and exit in upstream markets.

With appropriate density distribution of the niches of potential users a sequence of probit diffusion processes can lead to diffusion patterns which follow the traditional continuum of adoptions along an S-shaped profile.

Schumpeterian growth cycles

Growth retardation and, generally, Schumpeterian growth cycles can be explained in terms of the industrial dynamics within filieres – the system of vertical relations among users and producers of intermediary inputs. Vertical industrial dynamics is in fact ignited by the changes in the demand for production factors induced by the introduction of new technologies. The more biased the technological change, the larger the novelty of the new production factors with higher output elasticity, with respect to previous technologies, and the stronger are likely to be the effects on the upstream markets where the new products are supplied and exchanged.

This analysis is germane to the previous about the diffusion of new technologies.[3] The distinction lies in the focus of the analysis. We assume here that the new biased technology is in any case more effective than the previous or competing one. The positive effects in terms of production costs and related output growth can be distributed over time because of industrial dynamics in upstream markets and composition effects. After adoption in fact the changes in the relative prices, determined by the rates of entry in upstream industries, is likely to yield a stream of additional effects in terms of production costs and opportunities for output growth in global competitive markets. Out-of-equilibrium conditions determined by the horizontal effects of the introduction of technological change have a strong vertical effect in terms of a well-defined dynamic pattern in the distribution over time of the effects of technological change in terms of production costs and output growth.

The analysis of market dynamics in upstream markets now becomes crucial in order to understand the real effects of the new technology in downstream markets. Let us consider the market dynamics in upstream industries affected by the downstream introduction of new technologies.

The introduction of a new biased technology has an immediate and strong effect on the intermediary market for the newly highly productive factor in terms of an increase in its derived demand. With a given supply the first effect is a clear increase in the market price. If entry is not instantaneous, monopolistic rents emerge and the incumbent suppliers of the highly production factor can earn extra profits. If entry is barred by natural monopoly, actual costs disadvantages for newcomers due to reputation, inefficient size and limit pricing, incumbents can take advantage of high price cost margins and market prices can remain above competitive levels for a long period of time. Eventually, however, imitation can take place: large newcomers enter, often from adjacent markets. The market evolves from the original monopolistic setting to an oligopolistic one. Market prices are driven towards competitive levels: it is well known that the larger the number of players in an oligopolistic market, the closer are market prices to competitive levels costs.

Even if there are no barriers to entry, however, entry may be slow. Entry is the result of complex decision-making. Information about monopolistic rents takes time to spread, and in addition, the acquisition of the knowledge which is necessary to supply the new goods is time-consuming. The assessment of the costs and benefits of a new venture is also the result of lengthy process. Slowly and incrementally, however, market forces can adjust to shocks in the derived demand for production factors.

Only in the long term do extra profits attract newcomers, eager to benefit from the price-cost margins. Only then can the supply curve shift to the right. Price-cost margins shrink but equilibrium conditions are far away. The reduction of price-cost margins and the fall of prices in upstream markets has a direct bearing on the efficiency of downstream production processes because of the role of both absolute costs and the composition effects engendered by the changing levels of relative prices. The prices for the goods manufactured with the key inputs also decline and the equilibrium demand increase. The derived demand curve for the key production factor, however, also shifts to the right because of the positive effect on both output and production costs of the decline in prices. The new increase in the levels of the demand are accommodated by a new wave of entries only with further delays and price-cost margins may increase again.

Both absolute and relative prices are affected by such delays and out-of-equilibrium conditions. The cost conditions for downstream users are deeply affected both in terms of the sheer reduction of the absolute levels of market price for the essential inputs, and by the reduction of the relative prices. The reduction of relative prices engenders the composition effects, which directly affects the production costs in downstream industries.

The stylized analysis becomes far more complex when the chain of industries vertically related is longer and especially when technological change affects all the chain of industries. The interaction between horizontal and vertical effects becomes all the more intricate. The analysis of growth in industries where a biased technological change has been introduced and new intermediary inputs become relevant must pay great attention to the industrial dynamics engendered in upstream (and downstream) markets.

Retardation is likely to take place when adjustments are not instantaneous: the rates of entry in the sectors producing the more productive factors are not instantaneous as well as the rates of exit from sectors supplying now less productive inputs.

The telecommunications industry provides compelling evidence for the tight relationship between technological change, vertical industrial dynamics and total factor growth. The introduction of computer communication has magnified the demand for telecommunications services both from households and, especially, from firms. Firms rely more and more on computer communication to manage their production and marketing activities. The monopolistic set-up of the telecommunications industry prevented the reduction of prices made possible by the parallel introduction of digital technologies in switching, signaling and in transmission. Even after liberalization and privatization where state corporations were in charge of the supply of telecommunications services, entry has been slow and reluctant with high price-cost margins. All reductions in the prices for telecommunications services translated into the perceptible increase in demand for telecommunication services with strong output and substitution effects. A clear relationship between the levels of prices for telecommunication services and the efficiency of downstream production processes can be found in most advanced countries at the end of the twentieth century.

The distributions over time of the levels of the elasticity of entry to price-cost margins and of the derived demand to the price are crucial to assess the dynamics of general efficiency levels. Adjustments to equilibrium prices can be very long and require repeated and sequential entry (and exit). It is clear that a high elasticity of entry and exit to disequilibrium conditions is crucial for actual and potential productivity growth to match. Elasticity of entry to price-cost margins may be influenced by learning processes and information asymmetries. Similarly, the elasticity of demand to price is likely to be strong in the early stages when the use of the new production factor is small and the prices are high. Over time and with the decline of relative prices, however, such price elasticity declines.

As a consequence, the time paths, respectively of the change in the derived demand and of the change in supply, exhibit opposite trends. This affects the disequilibrium sequence of the short-term equilibrium market prices. This time path of price adjustments in upstream markets has strong effects, on the time profile of the actual growth of average costs and efficiency and the actual increase of output in downstream markets. Demand and supply interact with reciprocal feedbacks that shape the dynamics of the process.

The dynamics of such feedbacks between demand, supply, exit and entry can generate fascinating time paths of growth along a path, characterized by micro irregularities. Schumpeterian growth cycles can be explained in terms of the dynamic interplay between the sharp changes, both positive and negative, in the derived demand for production factors engendered by the introduction of new biased technologies, embodied in new intermediary products, the substitution effects of new inputs to old ones, the evolution of the market structures with increasing numbers of players on the supply and demand side and delayed exit and entry in the upstream markets for intermediate production factors.

Only the full appreciation of the composition effects makes in fact it possible to understand the reciprocal feedbacks between the elasticity of the derived demand and the elasticity of entry and exit. In turn it seems clear that industrial dynamics and specifically the evolution of industrial structures, because of the role of composition effects, play an important role in assessing the long-term evolution of the economic systems at the aggregate level.

Conclusions

The introduction of technological change has complex vertical effects in an industrial structure articulated in industries which provide each other with intermediary inputs and capital goods. Vertical effects have received little attention in industrial organization, much less than the analysis of the horizontal effects within each industry where competition is based on the introduction of rival products as well as new process, cost-reducing technologies.

Not only do vertical effects deserve more attention, but so too does the interplay between horizontal and vertical effects. The introduction of new technologies within each industry alters the competitive arena and as such leads to the creation of new barriers to entry and to exit, transient monopolistic rents and an array of quasi-losses and quasi-rents. The time span necessary for competitive conditions to be re-established depends on both the ability of competitors to imitate the new technology and enter the industry, and the capability of incumbents to adjust to the decline of their old market conditions. Upstream imitation lags and delayed exit have a direct bearing on the evolution of the absolute and relative prices for inter-mediary products and capital goods supplied to downstream users. Such delayed changes in relative and absolute factors costs in turn may induce firms in down-stream industries to change their technologies, either to introduce new technologies or to adopt new available technologies whose adoption was not yet profitable.

Within each industry in turn firms can react to the changing conditions of the market-place, engendered both by the direct introduction of innovations by rivals and competitors in the same product markets, and by the indirect effects of upstream dynamics when the differences in the behavior of upstream industries, across national and regional economic systems that compete in the global economy, matter. The new disequilibrium conditions, engendered both within the industry and across industries by the ripple effects of the introduction of new technologies, leads, because of the inducement mechanisms at work, to the introduction of new technologies.

The patterns of evolution at the system level can be explained by such interplay between horizontal and vertical effects. Adjustment to new technologies can only take place in the long term and in the context of disequilibrium where relative prices keep changing. The joint analysis of vertical and horizontal effects in an industrial matrix of industries also makes clear that the introduction of a new technology at a point in the system is likely to engender ripple effects which include the inducement to the introduction of further innovations in other industries and by other firms with a continual feed-back which constitutes the essence of innovation.

7 The dynamics of factors markets and technological change

Introduction

Structural change and technological change interact in many ways. The introduction and adoption of a new biased technology has a direct bearing on the demand for inputs and this in turn affects their market prices. New relative prices for basic inputs in turn change the cost conditions of firms, because of composition effects and hence induce firms to change their technology once again. A recursive relationship between the changes in the levels of relative prices and the introduction of contingent technological change takes place. All changes in the system of relative prices, due to non-technological factors such as demographic changes and emigration or immigration flows, also affect the cost conditions for firms and hence are likely to induce the introduction of new technologies. The rest of the chapter explores in detail the effects of the introduction of contingent technologies on the demand for labor on pages 94–97. Pages 97–101 focus the analysis on the discontinuities in the markets for production factors engendered by the overlapping of technologies which are not perfectly neutral and the risks of persistent macroeconomic disequilibrium to which economic systems are exposed. Finally pages 101–105 consider the effects of regional integration on the equilibria of local factors markets. This section shows the need for the introduction of new technologies in order to manage the effects of institutional changes such as the regional integration into a single market of countries originally characterized by important differences in their economic structure and especially in the levels of both absolute and relative factor costs.

Contingent technologies and the relative price for basic inputs

Factor prices are sensitive to the direction of technological change. The introduction of general technologies in fact has no effects on the relative price of inputs. General technologies consist only of a shift effect with no changes in the output elasticity in the production function with respect to the current factor intensity. Their introduction leads to a generalized increase in the efficiency of the production process with no asymmetric effects on the demand for each specific input.

The introduction of a new biased technology has the clear effect of reducing the derived demand for the now less productive input and increase the derived demand for the now more productive production factor.

The shape and specifically the slope of the supply schedule for production factors plays a crucial role in this context. When the supply of production factors is perfectly elastic the changes in the demand have no effects on their market prices. When instead their supply schedule has a finite elasticity the changes in the demand for production factors brought about by the introduction of contingent technological changes deserve careful examination.

Such effects are especially clear for basic inputs. Intermediary inputs are in fact themselves manufactured into the system and interindustrial mobility of production inputs can help, at least in the long term, to keep their prices constant and in any case in the proximity of equilibrium.

Let us assume that the slope of the schedule of the supply of both basic production factors is very rigid. All changes in the demand have a direct bearing on their prices. The introduction of labor-intensive technologies with no shift effects in a labor-abundant region is not likely to generate any actual increase in total factor productivity if it affects the factors markets in such a way that wages become equal if not higher than rental costs. The introduction of new labor-intensive technologies had the twin effect of increasing the demand for labor and decreasing the demand for capital. Unit costs of labor increased and unit costs for capital declined, resulting ultimately in undermining the very determinants for the apparent effects on total productivity levels generated by the introduction of such contingent technologies.

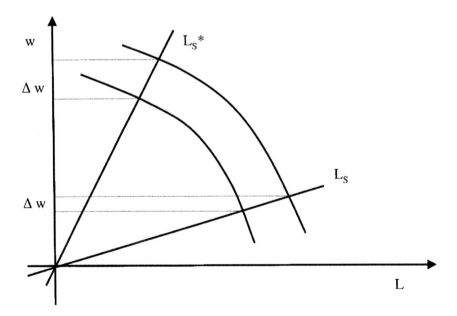

Figure 7.1 The supply and demand for inputs.

Labor markets for dedicated skills supply important evidence about such inter-actions. The supply of manpower trained in the wide range of skills related to information and communication technologies plays a key role in assessing the actual efficiency and the relative competitive advantage of countries after the introduction of new information and communication technologies. European countries lag behind in the timely provision of these skills with the ultimate effects of a steep and substantial increase in the relative wages which reach in absolute terms, levels which are even beyond the US average. This mismatch between the demand and the supply for competent manpower reduces, because of composition effects as well as for the absolute levels, the actual measured levels of total factor productivity levels associated with the use of information and communication technologies. A gap widens up between potential productivity levels and actual ones with negative effects on the competitivity of European products.

From this view point the provision of dedicated skill, as opposed to the overall levels of labor supply, cannot be regarded as the result of demographic trends, but rather the outcome of the levels of elasticity of the supply of training with respect to the demand. The analysis of the interplay between entry, supply and demand for intermediary and capital goods applies to the labor markets as well. In other words it becomes clear that the supply of appropriate skills should be regarded as endogenous to the economic and institutional system at large.

The institutional set-up for the mobility of labor and the eventual supply of skilled labor and specifically for labor force with formal training in universities and other higher education centers plays an important role in providing the proper amount of human capital to the economic system. Tenure and seniority systems may reduce the amount of workers willing to increase their levels of codified human capital. Tenure and seniority systems might perform better when learning by doing and informal on-the-job training are easily associated with long-term employment relations. High levels of interfirm mobility, coupled with flexible training systems, might perform better when new radical innovations, with high levels of require-ments in terms of formal training in codified knowledge, are relevant.

This is especially true in countries where education and training are provided by national and public institutions rather than by the market interactions of private firms. The timely adjustment of education programs and in general the flexibility of the higher education system to the directions and the rates of introduction of new technologies becomes a key issue to take a full advantage of the potential for increasing productivity levels, brought about by the introduction of new technologies which make use of new kinds of skills.

Similarly, it seems clear that because of immigration, labor supply can be considered, at least to some extent, endogenous to the economic system at large. The selective immigration of competent labor can help the long-term shift of the supply of dedicated skills. As such, selective immigration helps a country to reduce the mismatch between the demand for dedicated skills and the local supply with positive effects in terms of absolute and relative wages and hence the relative levels of increase of general efficiency measured in terms of production costs and of the total factor productivity of further waves of innovations with similar characteristics.[1]

An interesting new problem now arises concerning the ultimate viability of the introduction of sheer contingent technologies. The introduction of contingent technologies is viable only if and when the supply of the more productive factor generally exhibits low levels of price elasticities. Specifically the condition is clear: factors prices changes should be smaller than the change in the ratio of the marginal rates of substitution of the two technologies.

The elasticity of supply for production factors can be high when high levels of factor mobility are possible. In this case the actual position of the supply of the production factors that are used more can shift towards the right with the beneficial effect of the substantial accommodation of the increased levels of the demand and a substantial price stability. The institutions of labor markets, both within and between firms, play an important role in assessing the actual levels of elasticity of supply for skilled labor with proper levels of codified technological knowledge.

At the firm level, the choice between innovation strategies directed towards the introduction and adoption of general technologies and innovation strategies directed towards the introduction and adoption of contingent technologies is relevant. At the system level such an option seems severely limited by the actual conditions of factors markets. A research strategy directed towards the introduction of contingent technologies is viable only when abundant pockets of the more intensive and yet abundant factor are available.

The entry of new workers in labor markets either because of immigration or as a consequence of institutional changes in the social organization of families can provide such conditions. When the skill and the levels of human capital are relevant, similar effects can be generated by the introduction of radical institutional innovations in education, training systems and in immigration procedures and selection mechanisms.

Biased technological change and kinked derived demand

The economics of contingent technological change which is both locally progressive and locally regressive deserve much attention. In these conditions there can be one special value of the slope of the isocost line and hence of the ratio of wages to capital rental costs for which both technologies are equally productive. There is in fact an isocost line that is tangent to both isoquants. The choice of techniques and technologies for this slope is undetermined and can actually vary from one extreme to the other.

Depending on the value of the ratio of wages to capital rental costs, the labour-using and "inferior" technology may be superior and the rational producer will choose the inferior technology: the price efficiency is in fact much larger than the output efficiency of the more capital-using technology. For larger values of the slope and both higher wages and lower rental costs the labor-using technology will be abandoned and the new "superior" technology will be put to use.

This situation has many and important economic implications. First there is a technical interval defined in terms of input intensities or techniques which will not be used. This interval is exactly defined by the two tangency points of the same isocost with the two isoquants.

In Figure 4.3 this is clearly the case of equilibrium points A and B. In this situation the equilibrium demand for labor can be both L_A and L_B while the demand for capital can be both K_A and K_B. This effect is also clear when we observe the meta-maps of labor and capital productivity. They exhibit a concave kink for the values of either capital or labor comprised in the technical interval between the two techniques which belong to two different technologies. There is always a linear combination between the techniques A and B in Figure 4.3 (Stiglitz 1994).

Second, for that level of the slope the markets are unable to reconcile factor costs with factor intensities and hence choice of technologies. Some firms might retain the lower technology and others switch to the superior one. An important source of technological diversity among firms has been identified. The shape of the demand for both labor and capital now exhibits a double envelope with a long period of non-decreasing marginal productivity with respect to which a large variety of inputs can be accommodated in the production process.

Third, for small changes of the slope of the isocost, radical discontinuities with major economic effects will take place. For small increases in wages or decreases in capital rental costs, the demand for labor will drop suddenly. The demand for capital will surge dramatically. The adoption of the new technology will be sweeping. Total factor productivity at the system level will increase suddenly. Product price should eventually drop, because of the effects of total factor productivity.

Conversely we see that for small decreases in wages and/or increases in capital costs the reverse chain of radical changes will take place. The equilibrium demand for labor would increase dramatically as much as the demand for capital would drop while total factor productivity would actually decline and the product prices increase. Let us consider these four effects in turn.

When technological change is both locally progressive and regressive there is an important range of techniques and hence levels of input combinations which cannot be "rationally" practised. This has important implications in terms of significant discontinuities in the equilibrium demand for production factors. Such discontinuities are likely to make all adjustment processes in factor markets complex and slow. Moreover these discontinuities with fluctuations of the price of production factors risk to expose the systems to "bang bang" reactions with extreme swings in the demand for production factors.

When technological change is both locally progressive and regressive the schedule of the marginal productivity of labor acquires an interesting form with a kink in the function and a linear combination of factor intensities which keeps the levels of marginal productivity constant. The kink corresponds exactly to the range of techniques, defined in terms of input intensities, comprised between the two tangency solutions on the meta-isoquant map. All intersections with the labor supply schedule in that interval are meaningless and hence the system is exposed to the jump from the high levels of labor demand, below the "kink" and low levels of labor demand after the kink. Factor prices however do not suggest the one single correct technique to be used, neither do they push the firm to chose one technology.

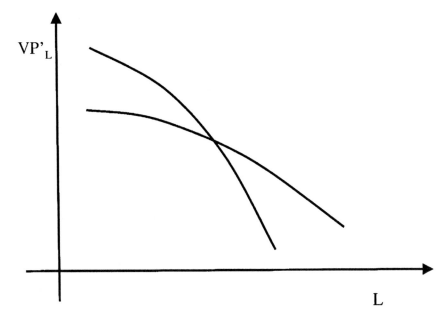

Figure 7.2 The kinked derived demand.

Exactly the same situation takes place in the markets for capital. For high levels of capital intensity, productivity jumps in a discrete way so as to determine a convex kink in the schedule again with symmetric effects in terms of a discontinuity in the equilibrium demand for capital from very low levels to very high ones. With labor and capital supply schedules characterized by some inertia and even "light irreversibilities", factor prices are likely to fluctuate between extremes as much as equilibrium quantities.

The rise of unemployment and the surge in real interest rates observed in most European countries between the end of the 1980s and mid 1990s can now be considered as being more directly related to the pace and direction of a locally progressive and regressive capital-augmenting technological change. This emerged as the result of a sequence of introduction of complementary and interdependent innovations characterized by substantial technical proximity, mainly shaped by the innovative efforts of fixed and human capital intensive US firms. These in turn were based upon learning processes in capital intensive techniques centered upon the newly emerging information and communication technological system.

The notion of local technological change has important implications from a macroeconomic viewpoint. The recent evolution of the European economy through the last decade of the twentieth century suggests that adjustments to technological transition can be complex. The relevant stylized facts are: (a) the new phase of technological change being introduced, shaped by the incremental and sequential rate of introduction and the capital-augmenting direction of the new technological system, being formed, mainly based upon new information and communication technologies; (b) the sharp increase of unemployment levels in most European countries, often followers in the technological race shaped by the US, advance in the introduction of incremental and sequential technological innovations of a highly capital-intensive nature; (c) the sharp increase in real interest rates and the strong demand for capital, with clear effects of credit rationing and competition between financial markets in securing the domestic demand for financial resources; (d) the segmentation of labor markets with a growing divide between the labor markets for skilled labor-force and the supply of low qualified, de-skilled workers; (e) the discontinuities in long-term dynamics of both productivity growth and in the trends of international convergence characterized by phases of decline in the actual, measured rates of growth of total factor productivity and the delay in the adoption of new technologies, often followed by sharp diffusion and catching-up in productivity levels.

These five stylized factors can be explained by the model of locally progressive and regressive technological change. The prices of products can drop, reflecting the increase in total factor productivity, only after a change in wages and capital rental costs. Prices decline after an increase in wages and the process is not smooth. Much empirical evidence as to the inverse relationship between prices and wages can be tentatively explained here. The adoption of the new "locally progressive and regressive" technology takes place only when general price efficiency does not balance output efficiency. In fact for such technologies adoption is not only a matter of output efficiency, but also of assessing price efficiency. Diffusion, that is the distribution of adoptions over time, is then guided by the changes in the ratio of wages to capital rental. Moreover diffusion can take place suddenly with a mass conversion to the new technology for critical values in the wage-capital rental slope. Reductions in the prices for capital goods embodying the new technologies as well as changes in wages can account for an S-shaped profile such as this in the diffusion process. Supply-side analysis of diffusion processes finds one more area of application at this point. All changes in the supply side of capital goods embodying new technologies can have major effects on the adoption of the technologies. The entry of new competitors in upstream markets and the fall of barriers to entry is likely to compress profit rates and reduce the prices for new capital goods.

This dynamic is likely to bring about changes in the slope of the isocosts and hence can induce a fast reorganization on the production process downstream with significant changes in the demand for labor and in the demand for the capital goods which enter a phase of accelerated diffusion.

When technological change is incremental and Hicks-non-neutral, a major gap can emerge between potential productivity growth and actual productivity growth. Actual total factor productivity growth will not take place as long as the new locally progressive technology in terms of output efficiency is also locally regressive in terms of actual price efficiency, i.e. with respect to the actual conditions of factor markets of prospective, potential users. Again we can reconcile the apparent paradox of the positive relationship between the increase in wages and the increase in total factor productivity. The increase in wages induces firms to abandon the old technology, to adopt the new and now superior one so as to gain advantage of the potential total factor productivity advance, which was still beyond the scope of prospective adopters.

The debate about international convergence towards homogeneous productivity levels assumes a smooth process, which in fact is likely, even when convergence actually takes place, to be characterized by strong irregularities with periods of fast catching-up followed by stagnation in productivity growth and even falls behind the standards of leaders (Baumol *et al.* 1994).

Once more it is clear that for small but critical changes in the value of the ratio of wages to capital rental costs, the system can suddenly switch to the new technology, adopt the new capital goods embodying the new technologies and finally, but only then, enjoy major increases in the overall levels of total factor productivity. Without such changes, total factor productivity growth can stagnate and new technology do not diffuse. When locally progressive and regressive combinations of the rate and direction characterize technological change, the economic system is exposed to significant forces that push it out of equilibrium. Only if the whole chain of standard reactions takes place at the same time, can the system remain in equilibrium conditions.

This, however, seems far too strong an assumption for the system to be actually able to behave accordingly. The reduction of prices, as induced by the increase in total factor productivity, is delayed by oligopolistic markets: as is well known, prices in industrial markets are downward sticky. Sticky prices impede equilibrium demand for the products to increase. Rigid demand levels delay adjustments of capital stocks to new desired levels. The reduction of employment instead is often obtained rapidly by firms. Hence aggregate demand is likely to decline, because of the reduction of solvable consumers.

The system is now likely to be characterised by a classical under-employment trap where the actual capital stock, already in place, is sufficient to meet the reduced levels of demand: hence investment is delayed. No compensation effects are likely to take place to reabsorb the employees who have been excluded from the firms adopting the new technologies: demand both for the products and for capital goods is rigid. Segmentation in the labor markets, due to the significant differences in human capital levels and generally in skills which are necessary to enter the more sophisticated labor markets associated with the new technologies, prevents the adjustments of aggregate labor supply.

In these conditions active macroeconomic policies oriented to sustain aggregate levels can help the system to exit from under-employment traps and regain full employment equilibrium conditions. Complementary labor policies which reduce labor market segmentation with appropriate reskilling interventions can help the system in the same direction.[2]

Technological variety and regional integration

New economic geography has recently drawn much attention to the economic effects of factor mobility, and specifically labor mobility, across regions. Krugman

(1991 and 1995) shows how increasing returns and externalities can engender centripetal forces leading to the concentration of economic activity in a few regions, when labor mobility is accounted for. Interestingly enough little attention has been paid in this literature to the effects of technological substitution and technological variety.

When technological variety is considered, economic integration of disparate regional systems can yield negative results in terms of reduced contingent technological efficiency. The hypothesis of technological variety contrasts sharply the current assumptions about a single best technology in place at each point in time. Technological variety is the result of the idiosyncratic and path-dependent processes of endogenous introduction of technologies that stress the productive role of locally abundant production factors.

Economic integration reduces the variance of local factor costs and hence affects the overall efficiency of the regions that use technologies biased in favor of locally abundant factors. Only high levels of dynamic efficiency and hence the rapid introduction of new technologies, able both to reflect the changing local endowments of production factors and to increase total factor productivity, can offset the negative effects of economic integration.[3]

The fast rates of international and regional integration within the European Union suggest that it is necessary to reconsider the basic assumptions of the standard economics of integration. The full exposure of all the European member states to the total mobility of labor, with the reduction of institutional idiosyncratic features of the local labor markets, and especially to the total mobility of the capital after monetary union and the rapid homogenization of capital markets, might have consequences not too dissimilar from the ones experienced after the unification of Italy in the nineteenth century. Central regions[4] can take advantage of the larger product and factor markets, while peripheral ones may discover a radical decline in their apparent efficiency.

When a single symmetric and exogenous technology is assumed, regional integration has no effects. These results, however, no longer hold when the hypothesis of endogenous technological variety substitutes the assumption of a single symmetric (exogenous) technology. A different context emerges when endogenous technological variety is accounted for: each region is able to elaborate its own technology according to the specific endowments and relative factor prices. At each point in time, not only do labor-abundant countries use more labor-intensive techniques, but also such techniques belong to more labor-intensive technologies. By the same token capital-abundant countries should use more capital-intensive techniques and technologies. Technological variety across local systems is now a natural outcome of this line of analysis. Provided systems differ in terms of factor endowments, tastes and hence product markets.

Let us consider for the sake of clarity a simple case with two regions and two technologies. Region A(merica) is characterized by labor scarcity while natural resources are abundant. Wages are high as well as capital intensity. The technology in place, as it emerged from centuries of biased learning aligned along a technological path characterized by the original capital-intensive techniques, is capital-deepening as well. Region B(ritain) is characterized by labor abundance and relative scarcity of capital. Wages are lower as well as capital intensity. The marginal productivity of capital, however, is also lower. Region B has in fact been able to capitalize upon localized learning processes and has improved its labor-intensive techniques over time, shaping a new different technology. Both factor markets and technologies differ across the two countries. Factor mobility enhanced by product mobility should push cheap labor from labor-abundant countries towards capital-abundant ones. Capital also should flow towards countries where it is scarce in search of higher rental costs. Wages and rental costs should converge.

For countries of similar size, product in both countries should decline and average costs increase. Labor is in fact subtracted from labor-intensive production as well as capital being taken away from capital-intensive ones. The higher costs of both factors suggests there is a need to increase labor intensity in countries where capital-intensity technologies are in place and to augment the capital intensity where labor-intensive technologies are at work. When the two countries differ in size and the flows of respectively capital and labor across the border are not important for one country, but significant for the smaller one, the outcome of regional integration is even worse at least for the smaller country. Labor leaves the country B and it is fully and easily absorbed by the labor market of country A with no effects on A(merican) wages. Wages however increase substantially in B(ritain). On the other hand, the inflow of capital in B has significant effects on local rental costs but trivial ones in country A. Rental costs and wages are not altered in country A while they are radically changed in country B: they are quickly driven by mobility to the A(merican) ratios. Country A does not suffer from economic integration while country B is put under strain. Country B now appears to be less efficient than country A. Yet previous equilibrium conditions have been altered by integration. Without integration both countries were equally efficient.

These negative consequences can be overcome only if regions do not keep their original technologies, but try and change them according to the new factor market conditions. In the symmetric case when two regions of equal size are concerned both are induced to change their technology. Not only do factor costs converge but also technologies. Clearly, a symmetric production function, where the output elasticity of capital equals the output elasticity of labor at obvious 0.5 levels, is the outcome of this process. When the regions differ in size, however, only the smaller ones must carry the burden of adjusting their technology. The asymmetry here is even worse. On the one hand agents in country B can imitate the technology of country A, instead of generating new technologies from scratch. On the other however their competitive conditions are put in jeopardy by the delays and lags with respects to innovators in country A.

In a context of technological variety, the economic integration of disparate regions jeopardises their static efficiency. When the standard assumptions about one single general technology are released and more reliable hypotheses are made about technological change, regional integration, in the form of increased mobility of labor and capital, can have very negative effects on aggregate welfare. Only high levels of dynamic efficiency can overcome the problems arising from the economic integration of disparate regions. Dynamic efficiency consists in the capacity to generate timely new technologies that are appropriate to the changing conditions of both products and factors markets,

The new thinking in the economics of innovation portrays technology as endogenous and putty. Technology is the result of the path-dependent and localized efforts of firms facing the specific constraints of local product and factor markets. Within each region, characterized by a low variance of the relative prices of both products and markets and hence of natural and historic endowments, a specific and localized technology emerges, one which is able to make the best possible use of production factors locally abundant and hence to save on locally scarce ones. This approach leads us to consider the hypothesis of a great variety and coexistence of technologies in place at each point in time in the general markets.

Only technological change can overcome the negative effects of regional integration. When and if new technologies strongly biased in favor of the new conditions of the factor markets can be introduced in a timely fashion, the decline in apparent efficiency can be reversed. In the process, however, smaller countries are likely to suffer for both their lower apparent productivity and for the relevant efforts that are necessary to change their technology.[5]

A strong policy issue emerges from this analysis. Countries that are considering increasing their integration and reducing the barriers to full mobility of capital and labor should be ready to face a decline in their short-term apparent efficiency, especially when and if their technology is strongly biased in favor of their abundant production factor.

In these conditions it is clear that a strong innovation and technology policy should accompany and sustain regional integration. Such an innovation policy is necessary to help the drastic change in the technology now required and to reduce the time window that is needed to change the technology. The aims of such a complementary technology policy are clearly twofold: to help firms to increase total factor productivity levels and to change the mix of the relative efficiency of the different production factors, following the new relative factors prices and hence biases as they have been defined in the course of the integration process.

Strong efforts to favor the technological capabilities of peripheral regions seem necessary when regional integration takes place and factors mobility can change the traditional structure of relative prices. From an innovation policy viewpoint this suggests that, after integration, peripheral regions should receive a large portion of public funds available for research activities.

The new evidence on the strong regional concentration of innovation activities and increasing returns in the production of technological knowledge is an important result of the large amount of empirical literature gathered in recent years. Innovation activities cluster in a few regions world-wide and in a few technologies. Increasing returns in the production of technological knowledge take place within technological districts and technological clusters where qualified interactions among connected innovators make it possible to take advantage of the modular indivisibility and cumulativeness of technological knowledge. In this context the notions of innovation systems, irreversibility and collective knowledge become relevant. Learning and technological communication are key factors in the definition of the rate and direction of technological change within technological systems.

Technological systems such as technological districts and technological clusters can be identified within a theoretical framework which values local externalities, irreversibility and endogenous structural change (Antonelli 2001a).

Regions and local innovation systems, characterized by high levels of superfixed production factors and technological complementarities, business turbulence and conducive conditions for technological communication are likely to experience fast rates of introduction of technological changes. Location within well-defined regions is an important factor favoring the generation of technological knowledge in highly productive conditions. High levels of regional concentration of innovation capabilities within technological districts play a major role in fostering the rate of introduction of technological change at the country level. The understanding of increasing returns in the production of technological, however, has important and conflicting implications from an innovation policy viewpoint. The regional concentration of public resources available for technological and scientific research in a few technological districts seems appropriate in order to increase the efficiency of the production of knowledge and the rate of introduction of new technologies.

On the other hand our analysis of the effects of regional integration in terms of the consequences for the local endowments of production factors and hence on the relative fitness of the local technological specificity has shown how important it is for peripheral regions exposed to regional integration to change their technologies. Technological change is the most important tool peripheral countries can use to overcome the emerging problems raised by the homogenization of factors markets. From this viewpoint all the public resources available to increase the rates of introduction of new technologies should be concentrated in peripheral regions.

An important divide here takes place between the centripetal implications of localized increasing returns in the production of knowledge, and the centrifugal need to concentrate most technological efforts in peripheral regions which are rarely characterized by a strong technological and scientific infrastructure.

Conclusions

Technological change can only be understood when the analysis is embedded into the recursive relationship of a systemic approach. Irreversible investments, and decisions, at each point in time, are made upon the expectations of myopic agents, unable to foresee the future. At each point in time technological change is induced by changes in the structure of relative prices which stimulate modification of the routines of the firms, because they engender a mismatch between expectations and the actual conditions of the product and factor markets. Irreversibility limits the possibility of facing unexpected events with standard quantity and price adjustments. Innovation is then induced to solve the new emerging problems. The interplay between the changes in the supply of inputs and in the demand for the products of the firm, irreversibility and the rate and direction of technological change is strong and effective and provides the basic framework of analysis into which the long-term direction of growth of an economic system can be grasped. The analysis of the interaction between technological change and structural change can only be conducted in a framework which appreciates the disequilibrium condition into which feedbacks and ripple effects keep diverting agents from expected equilibrium points.

8   Product innovation and barriers to entry

Introduction

Heterogeneity of factors markets plays a key role in our analysis so far. Production costs can vary across heterogeneous factors markets because of differences in endowments and hence in relative factors prices, even when firms are able to adopt most advanced technologies, if and when a mismatch emerges between the bias of each factor in the production function and its relative costs.

In a single economic system where single prices for homogeneous production factors apply, such effects would not take place. In a complex economic system where heterogeneous factors markets exist, technological heterogeneity can also arise. Actually technological variety should compensate for factors markets heterogeneity.

Potential adopters and incremental innovators of general purpose technologies which exhibit a strong bias in favor of locally scarce and therefore relatively more expensive inputs face significant cost asymmetries with respect to adopters located in countries where the bias is consistent with the local endowments. By the same token innovation strategies should be localized, that is, be consistent with the endowments and hence the relative costs of most productive inputs.

A serious problem arises when technological change is not able to compensate for factors market heterogeneity and a gap widens between the levels of potential total factor productivity and actual ones.

Heterogeneity among consumers, however, can play an important role in assessing the working of a complex economic system, one where not only the local endowments for production factors differ, but where the tastes and preferences of consumers cannot be reconciled with the traditional notion of a representative consumer. Groups of consumers with distinctive tastes, influenced by history and traditions, culture and institutions, coexist in a complex economic system.

Niche strategy can provide an opportunity for firms located in less favored countries to cope with a general purpose technology which makes extensive use of locally rare inputs. A niche strategy consists in an innovation strategy able to focus not only the adoption of a new process technology, but to bundle it with the introduction of product innovations with strong idiosyncratic characters which can have high prices in the market place.

Barriers to entry and monopolistic competition

Product competition rather than price competition is relevant when firms are able to change not only their process technology, but also can introduce product innovations and consumers are heterogeneous in terms of preferences.

When product rather than prices competition takes place, the market prices for the goods are not the single vectors of all relevant information. The distribution of tastes and needs among heterogeneous consumers becomes also an important factor assessing the working of competition in the market place. Firms can take into account the effects of such consumers' heterogeneity when they choose their market conduct.

The introduction of new products, manufactured with new process technologies, can compensate for the cost asymmetries engendered by high relative prices for most productive inputs, with high levels of customer preference. Such products are replaced with less expensive products only with some reluctance by customers. Customers take into account not only the levels of momentary market prices but also the specific features of the new products. The levels of specificity of such products are such that they cannot be considered proxies for competing products, but rather well-identified goods which deliver a bundle of unique services.

Niche product innovation consists in the introduction of a new idiosyncratic product. The new product provides customers with higher and better services with respect to other existing products. Such products cannot fully replace existing products, but cannot be, in turn, fully replaced by new generations of products. These products survive in specialized niches for a smaller range of customers and a smaller bundle of services. Each niche product has a well-defined identity with high levels of idiosyncratic use. The reputation of producers and their brand help to increase the idiosyncratic levels of dedicated usage.

Niche product innovations increase the variety of products in the market place and the range of choices for prospective customers. In this context it seems clear that the effects of product innovations in the market place cannot be analyzed within the context of standard "perfect" competition models, but rather within the well-established framework of monopolistic competition (Momigliano 1975).

The introduction of a new product has a direct effect on the demand for existing products. The demand for existing products shrinks when a new superior and less expensive product is introduced.

Hedonic prices can be considered here a useful device to take into account jointly the levels of quality and the monetary cost for the consumer of a new product. Hedonic prices do contribute to assessing the effects of the quality of products. From this viewpoint they represent an important step with respect to the analyses based upon monetary prices especially for intermediary products where an objective assessment of the quality of a good can reach universal consensus. The hedonic content of a final product, mainly if not exclusively destined for household use, however, requires the assumption of a representative consumer. The problem of heterogeneity in consumers' tastes and preferences is not resolved, but even worsened when hedonic price methodology is used. A variety of hedonic indexes would become necessary, for the same product, according to the variety of consumers or groups of consumers and their respective density.

The problem is not the objective quality levels of products but rather the quality perceived by heterogeneous consumers and the demographic distribution of such tastes: i.e. how individuals are spread in a space of product characteristics. The match between the features of a product and the characteristics of the utility functions of an array of individuals become the key factor to assessing the actual quality of a product and whether it can be substituted, for given market prices, to another one, and by how many consumers.

In this context cross-demand elasticity to prices appears relevant. For any given level of the prices of each product the demand levels change when the prices and the perceived quality of other products change. Customers substitute the old product with the new one and hence reduce the demand for the old product. Cross-demand elasticity to prices, however, is infinite only for homogeneous products. Cross-demand elasticity to prices is less than infinite when the new product has a distinctive character and matches the idiosyncratic needs and tastes of a well-defined group of consumers. The old product survives in the market place, next to it, because of its distinctive features. The values of cross-demand elasticities to prices among chains of products vary according to the varying levels of substitution across individuals. When the substitution is nearly perfect, the cross-demand elasticity to prices approaches infinity. When the substitution is less than perfect, the cross-demand elasticity to prices is less than infinite.

The potential market for a new product depends on the actual scope for substitution with respect to other existing products. In turn the scope for substitution depends not only on the differences among products in terms of monetary price, but also (and mainly) on the distribution of preferences of consumers across the now wider range of products, and, most importantly, upon the density of the groups of similar consumers.

The distinction between heterogeneous and homogeneous product competition has important consequences on the analysis of entry. In perfect Marshallian competition entry is induced by the extra profits of incumbents and lasts until the decline of the market prices reaches the minimum average costs of the most efficient firms. The products delivered in the market place by the newcomers are perfectly homogeneous and no distinction can be made between the new products and the old ones.

Monopolistic competition, on the other hand, emphasizes the limited substitution among heterogeneous products and assumes that newcomers are attracted by the extra profits of incumbents and enter in the market-place. The demand for the new products reduces the levels of the demand for incumbents. Entry takes place until the demand is tangent to the average cost curves of incumbents.

Entry can be limited by the strategic behavior of efficient incumbents who fix prices taking into account the cost conditions of potential competitors. Provided that a cost difference exists between incumbents and potential competitors, incumbents can limit the entry of less efficient competitors by means of limit prices. Standard textbook analysis of barriers to entry suggests that incumbents fix limit prices in the proximity of the production costs of potential competitors for the same homogeneous product. Limit prices are then applied to a single product demand (Sylos Labini 1956/1962).

No distinction is made between the demand for the product sold by the incumbent(s) and the product sold by the potential competitor(s). Consumers are not supposed to be able to distinguish among them. Barriers to entry apply to the industry, rather than to the firm.

Great attention has been conveyed by the elegant contribution of Modigliani (1958) on the effects of the profits of incumbents on the price elasticity of the demand curve. The larger the slope of the demand, the smaller is the direct price elasticity, and the larger is the limit price and hence the mark-up for incumbents.

When the analysis of barriers to entry is developed in the context of monopolistic competition and product heterogeneity is taken into account, a new framework emerges.

Incumbents are aware of the effects of the production and sale of other similar, but not homogeneous products, to a heterogeneous population of individuals with their own specific tastes and preferences, on the demand for their own good. Such effects are measured by cross-demand elasticities to prices. A family of negatively sloped demand curves replaces the single demand curve for the homogeneous product of the industry. For each price level of products that can be replaced by their own product, incumbents face a specific down-sloped demand curve. When rival products have no effects whatsoever on the demand for the product, the demand lies far on the right. Alternatively, when the effects of the introduction of similar products are very strong, the demand curve for the product sold by the incumbent lies at the minimum level of the range of demand curves.

The level of the specific demand for each firm depends on the actual substitution between its own product and the substitutes. As a matter of fact the actual position of the demand curve for each product, out of the bundle of demand curves, depends on the actual substitution effects at work with other products which in turn are a function of both the prices of rival products, the levels of their relative preference with respect to existing ones and once more the distribution of tastes and preference in a population of individuals which differ with respect to the specification of their utility function.

Network externalities moreover can apply in this context so that the introduction of a new product has not only the standard negative effects on the position of the demand if marginal fringes of consumers substitute the new product for the old one, but also positive ones when the increased size of the stocks and flows of purchases for the same class of products affects the preferences of the consumers for the original product as well.

The demand for a homogeneous product is immediately affected by all changes in the market prices for any additional quantity of the identical product brought to the market-place. Cross-elasticity is infinite. The demand for a product that has no substitutes whatever is not influenced by any changes in the market prices for other products at large. Cross-elasticity is zero. The single producer of a product with a cross-elasticity equal to zero enjoys a full monopoly and can extract the maximum of quasi-rents from the production and sale of the product.

In between these two extremes a wide range of intermediary possibilities exists. Such a range deserves a careful examination.

Niche-pricing: a simple analytical exposition

As is well known, the demand curve for a product is derived from the constrained maximization of the utility function for a given level of revenue and product prices, with well-identified levels of preferences for each product and the horizontal sum of the individual demand curves. The simple derivation of the demand curve, however, misses not only the revenue effects, but also the compensated substitution effects. The former revenue effects can be taken into account by the Slutsky equation. The compensated substitution effects are rarely considered. Even less considered is the issue of the heterogeneity of tastes of consumers when choosing among different products.

Changes in the prices for any products in the utility function have direct effects on the demand for a given product, according to the ratio of the marginal utilities. The demand for a product with high levels of preferences will be affected only to a limited extent by the changes in the prices of other products, especially after compensating for the revenue effects. The demand for a product with low levels of preferences instead will be strongly influenced by the changes in the prices of the products that can be substituted for it.

By the same token it is clear that the introduction of a new product with high levels of preferences will have strong effects on the demand for the existing products that can be substituted. The introduction of a product innovation which combines high levels of preferences for a large group of customers and low prices can have strong effects on the demand for substitutable products.

According to this line of analysis the standard specification of the demand curve, derived from Cobb-Douglas utility functions, seems more appropriate for analyzing markets for homogeneous products rather than markets for heterogeneous and yet partly substitutable products. Ordinary specification of the demand equation reflects only the direct price elasticities and, even when the revenue compensation effects are considered, does not fully account for the indirect effects of the levels of the prices for substitutes, that is, does not make explicit the role of cross-elasticities. As such it seems inadequate to analyze a market characterized by the continual introduction of a variety of heterogeneous products. Traditional demand curves are not able to reflect the effects of the introduction of product innovations.

An effort to tackle the effects of actual cross-elasticities is necessary in order to achieve the integration of the effects of the market prices of substitutable products into the demand curve for each product. A preliminary attempt can be made along the following lines.

In an ordinary demand function, derived from standard (Cobb-Douglas) utility functions, the effects of the preferences for substitutes upon the demand for each product is measured by the ratio between the parameters which measure the relative preference assigned to each good by the consumer.[1] The number of such consumers is not considered.

A utility function which makes it possible to make explicit the cross-substitution effects of the changes in the prices for substitutes becomes necessary. Henderson and Quandt (1971/1980) provide the following simple example for a two-goods case:

(3) $U = Q_x Q_y - (Q_x)^2$

Constrained maximization subject to the budget (R) yields the following demand function:[2]

(4) $Q_x = (R - P_y) / 2 P_x$

In equation (4) the demand for the good x is influenced both by the price for the good x and by the price for the good y.

The linear transformation of the demand equation can parallel these changes in the utility functions. Let us assume that a standard linear demand curve is provided in equation (5):

(5) $P_x = A - b Q_x$

The intercept of the demand curve A can now be considered under the influence of the effects of the prices (P_y) of the other substitutable product y on the willingness of customers to purchase the product x. Hence we can write:

(6) $A = A' - 1/(P_y)^\gamma$

In equation (5) γ^3 measures both the effects of the prices of other partial substitutes and their relative preference on the demand for the good considered.

The substitution of equation (6) into equation (5) yields a linear specification of the demand curve which is able to take into account the variety of partly substitutable products in the market-place on the demand for each product:

(7) $P_x = (A' - 1/ (P_y)^\gamma) - b Q_x$

The market conduct of each firm can now be set with respect to its own market niche. The size of each market niche can be defined only after the effects of partial substitution have been taken into account. Cross-elasticities across products, together with the ratio of marginal utilities of partly substitutable products and the composition and density of consumers in the space of product characteristics, can contribute the understanding of the actual levels of γ.

On these bases the basic insight of barriers to entry can be retained: firms choose their own market strategy strategically. Each firm fixes its own price and the characteristics of its product taking into account the effects of the conduct of actual and potential rivals and the reactions of a variety of consumers. Each firm tries and assesses the effects on its own product market share of the general demand curve

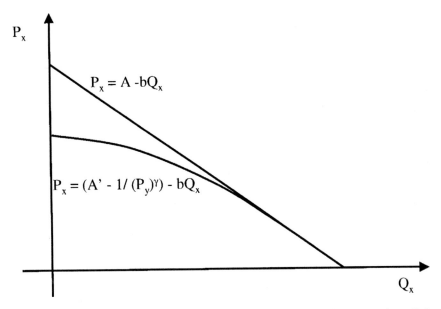

Figure 8.1 The demand curve when the effects of product substitution are made explicit.

for that class of substitutes, that is on their own individual demand curve, of the conduct of rivals and specifically of the costs and perceived quality of their products. The identification of the actual market for the distinct product of each firm requires a careful assessment of the distribution of tastes across consumers, weighted in terms of demographic density, and the cost conditions, competence and technology of rivals.

This approach assumes that entry can take place at any time, albeit in fractional parts of the general demand curve. The Sylos Postulate does not hold, not only because of the strategic interaction between incumbents and newcomers in terms of the *ex post* effects of entry in homogeneous products markets, but also and mainly because of the heterogeneity of consumers and the scope for identifying and targeting groups of idiosyncratic consumers, of varying size, with the introduction of new products. Product innovation and product competition, as opposed to price competition in standardized products markets, are the basic assumptions here.

This amounts to assuming that contestability of specific product markets is less than perfect. According to Baumol (1982: 3 and 4) "A contestable market is one into which entry is absolutely free and exit is absolutely costless . . . the entrant suffers no disadvantage in terms of production technique or perceived quality relative to the incumbent." Specific product markets are contestable only to a certain extent: the quality and the features of the products of newcomers are in fact different, not necessarily lower or higher, than the quality of the products already sold in the market-place.

When rivals have far better cost conditions, their market prices can be low and hence even for less attractive goods the number of loyal consumers ready to confirm the purchase of the products delivered by the incumbent can be too low to cover average costs. Alternatively, however, it is now possible to consider the possibility for newcomers to enter even if the costs conditions of the incumbent(s) for direct substitutes are lower. This is possible when the newcomers are able to identify a group of consumers who are willing to pay for products that are specifically designed for their own tastes and needs. This is all easier when existing prices are set above standard limit prices for homogeneous products.

The concept of barriers to entry is a powerful analytical tool by which to understand the oligopolistic strategic interaction among firms with different cost structures. Its use, however, has been mainly developed in a context of homogeneous product competition. Its application to a fully explicit context of monopolistic competition reveals new elements of interest. In a monopolistic competition context each firm produces a distinctive product with strong elements of identity and reputation and generates a different bundles of services for their users. As such each product is perceived by consumers as an imperfect substitute for other goods.

In this context the notion of industry, defined as a group of firms which sell the same identical good, loses its scope of application. The same is true for the notion of a general demand curve for homogeneous products. Each firm has its own individual demand curve. The size of such individual demand curves depends upon the ability of each firm to deliver products that have a distinctive utility for as large a group of satisfied consumers as possible. In turn the variety of consumers, in terms of tastes and needs, becomes a key assumption of this line of analysis. Clearly with a "representative" consumer the variety of heterogeneous products would collapse into standard homogeneous ones.

The intrinsic heterogeneity of tastes and preferences across consumers provides the basic barrier to entry in each specific product market. Rivals cannot produce an identical good that perfectly substitutes for any other product. Rivals, however, can produce goods which attract large portions of the group of original consumers of any other good. The oligopolistic race among firms hence is based upon the definition of the size of each individual market.

Here the key role of product innovation shows its effects. The size of the individual market for each product will depend not only upon the market price but also and to a large extent upon the number of consumers for which the levels of the relative preference for each product is higher with respect to other substitutes. Brand loyalty, visibility and reputation clearly help in strengthening the levels of relative preference for each distinctive good.

Each producer can be considered a niche monopolist. Each niche monopolist will define a monopolistic price in its own (small) market niche. Such a market price can be termed niche-pricing. Niche-prices can be higher than traditional limit-prices, provided the slice of market demand for their own distinct products, after considering the production costs and hence the market prices for rival products, has been taken into account, as well as the levels of preferences that customers are ready to attach to each rival product. Standard limit prices, that is prices which are fixed to the levels of the costs of rival competitors that produce homogeneous products, can be higher if and when the effects of cross-elasticities and the ratio of marginal utilities on the intercept of the standard demand curve are high. Niche-prices will be equal to full monopolistic prices only when cross-elasticities are zero and the demand for each product is not affected at all by any other product.

Niche-prices for new products that are able to match high levels of preferences for a well-defined group of customers, and as such yield only a limited fraction of the total potential demand to rival goods, can be considerably much higher than competitive prices. A large mark-up can be incorporated into such prices.

The difference in costs between actual and potential rivals of course plays an important role. The entry of new competitors with low production costs can reduce the individual demand curve of the incumbent to levels that are even below the average cost curve and force the exit of the firm, even if a small fraction of loyal and satisfied consumers would be happy to keep on purchasing the good delivered by the incumbent.

Niche-prices can be considered the end result of a strategy of dynamic limit pricing where incumbents maximize the discounted flow of profits, by fixing prices that are higher then actual limit prices. Incumbents in this case assume that entry is going to take place, but with a delay. With respect to the literature on dynamic limit pricing, however, the features of each product remain unique (Gaskins 1971). Product rather than price competition, however, applies and entry cannot fully eradicate the demand for each specific product.

With respect to the static limit pricing literature, two issues become relevant. First, niche-pricing seems able to solve a basic ambiguity in traditional limit-pricing analysis as to the number of incumbents. Traditional limit pricing analysis rarely approaches the problems arising when incumbents are more than one. When there are many incumbents the strategic interactions among them become complex and the definition of a single limit price can become problematic. Only a strong assumption about collusion among incumbents can provide a straightforward solution. In our approach instead each firm is incumbent in its own product market.

Second, when niche-prices matter, two price elasticities affect the profits of incumbents rather than one: the direct elasticity to the price of the product and the cross-elasticity to the price and preferences addressed by other products that can be partly substituted. The lower the cross-demand elasticity to the prices and the perceived quality of other products, the higher are the rents each firm can extract from its own product. The higher the direct demand elasticity to the own price of the product, the higher are the price-cost margins. When the slope of the demand curve in fact is small (in absolute terms) the equilibrium demand is larger and, for a given average cost, the levels of profits are larger too.

The introduction of new substitutes can actually have positive effects on the demand for each product when network externalities apply. The introduction of new and cheaper products that attract new groups of customers can reduce the demand for the existing good because marginal fringes of consumers substitute the new product for the existing one, but at the same time it can increase the utility levels for other users. The larger stock of compatible and interactive products can have positive effects on the services delivered by original products to their consumers. The algebraic effect on the position of the demand for the specific good of the entry of new products in the market-place, even if they are substitutes rather than complements, may be positive, rather than negative. Product competition in new information and communication products provides larger empirical evidence on the strong effects of network externalities on the demand side and confirms the role of niche strategies and especially product innovation as the single viable market strategy for firms.

Our analysis provides not only a theory about the conduct of incumbents in niche-product markets, but also a theory of entry. Newcomers can enter a market, even when their production costs are high, provided they are able to identify a niche and target their innovation strategy towards a well-selected group of consumers. Important implications in terms of division of labor can be derived. Firms located in regions where the relative prices of production factors match the specific requirements of a given technology in terms of output elasticities can target mass markets. On the other hand, firms located in countries where the mismatch between the features of a given technology and the local factors endowments exists, should try and enter only niche-product markets.

The adopters of general purpose but biased technologies with a wide range of applications and a distribution of output elasticities which is not aligned with the local endowment of production factors can face major cost asymmetries in the global markets and survive in the market-place only if they are able to introduce well-targeted and unique product innovations, taking into advantage consumers' heterogeneity. Without the stream of income generated by such product innovations their total factor productivity after adoption would be higher with respect to previous technologies, but in any event lower with respect to firms located in countries where most productive inputs have lower relative prices.

Firms located in countries where relative prices of production factors are not aligned with the distribution of output elasticities could not delay the adoption and the incremental innovation of the new technologies: the new technology is in any case superior to the existing ones. Any delay would imply a clear opportunity cost. Such firms would still be less cost-effective than their competitors based in countries with a more favorable distribution of endowments.

Conclusions

Heterogeneity of consumers plays an important role in the analysis of industrial dynamics. When the representative consumer is replaced by a variety of consumers with different tastes and preferences, product competition rather than price competition is relevant. The introduction of product innovation able to target well-defined product-niches and to match the specific requirements of groups of consumers can secure a fraction of the general demand curve for well-identified products and generate relevant mark-ups.

Firms able to introduce product innovations can fix niche-prices which take into account the cross-demand elasticity for each product with respect to the prices and the characteristics of the other products, especially if they are brought to the market-place by the entry of new competitors. Product markets are contestable, but only to a limited extent when products are not homogeneous. The determinants of the process of substitution among products by consumers becomes a relevant issue in understanding the strategic behavior of firms.

The prices of the substitutes play a role in such a process. A much more important role, however, is played by the ability of each product to match the specific tastes and preferences of groups of consumers of different size. The competence of firms, in terms of their ability to generate product innovations and to build reputation, induce strong consumer loyalty and hence reduce the risks of substitution. Low levels of substitution make it possible to increase the levels of niche-prices.

In turn niche-prices and consequent mark-ups can balance the cost asymmetries emerging from the mismatch between the bias in technologies and the local endowments. Less effective firms can survive, and a variety of profit rates together with technological variety are the ultimate result of the integration of both factors markets and the heterogeneity of consumers' tastes.

The introduction of niche-product innovations and/or of process innovations directed towards generating properly biased technologies, able to minimize the mismatch between local endowments and output elasticities of production factors, become the two horns of a locally aware technology strategy.

9   Relative prices and international competition in the global economy

Introduction

International trade has symmetric and positive effects for all parties when a clear technological variety exists. Each country specializes in the technology which makes the best usage of locally abundant production factors. Following Ricardo, international trade is beneficial when the technology for the production of wine differs from the technology for the production of cotton. Specifically wine technology is successfully applied in Portugal as it makes intensive use of locally abundant resources, while cotton technology makes more intensive use of factors, such as skills and capital, relatively more abundant in England.

Technological variety is assumed as a founding condition for international trade to take place with mutual benefit for parties engaged. The performances on international markets of each economic system are strongly influenced by the structure of relative prices together with the specific bias of the technology in use. Much attention has been paid to the role of absolute factor costs and to the rate of technological change in assessing the performances of each country in the international market-place. Very little attention has been paid to the role of relative prices and to the direction of technological change. Yet it is clear that relative prices interact with absolute factor costs, the bias in the direction of technological change and the levels of actual total factor productivity in assessing the average costs of products. Countries with lower average costs are able to appropriate larger markets shares and possibly to earn extra profits especially if less efficient competitors delay exit.

With a proper bias in the direction of technological change, relative prices can actually compensate for absolute factor costs and lower total factor productivity associated with the shift effect. With a given single, non-symmetric, general technology, countries where the production factor which is more productive is also the most abundant experience lower average costs and hence acquire larger market shares on international markets.

When technological variety is taken into account it is clear that the direction of technological change practised in each system becomes a central issue. Countries able to implement their technologies along technological paths shaped by the relative abundance of a specific production factor can enjoy significant competitive advantages. Such countries can even have higher absolute factor costs, provided they are compensated not only by the levels of total factor productivity, but also, and even mainly, by the structure of their relative prices.

Finally, countries which are not able to innovate and generate their own localized technologies can take advantage of imported technological innovations and have a strong competitive advantage on international markets, provided they are able to shape the structure of their relative prices as close as possible to that of innovative and best-performing countries.

Relative prices are a major factor of pecuniary externalities. Firms with a low purchasing power, because of high levels of absolute factor costs, located in a country where the most productive factor is cheapest and the least productive input is most expensive, can successfully compete with firms based in countries with lower absolute costs but the "wrong" ratio of factors costs.

It is clear that in a dynamic context of analysis, the international competitivity of each economic system in global markets depends on the correct co-evolution of both the structure of the system, in terms of factors absolute and relative prices, and the technology. Countries able to generate new technologies that make a more productive use of locally more abundant resources and are able to further reduce their production prices and increase their supply can experience very fast rates of economic growth.

Here the distinction between contingent and general technological change plays a major role. The distribution of the gains from international trade are likely to be substantially symmetric only when a new general purpose technology, characterized by a significant shift effect and neutral both in the country of origin and in the country of adoption, is introduced. In this case all regions have a clear incentive to adopt the new technology. If all countries are equally able to take advantage of the new technology there is no alteration in the distribution of the gains from international trade. Transient monopolistic power associated with diffusion lags based upon information asymmetries lasts only in the very short term.

In all the other cases, that is both when the new technology is contingent and the new technology is only locally neutral, technological change has asymmetric effects in the trade among heterogeneous countries.

The effects of the introduction of general technologies

At a time of the introduction of new radical technologies that reflect the original endowments of production factors in innovating countries, but induce a factor intensity different from the one currently in use in adopting countries, international competition does not have positive effects for imitating countries, at least in the short term. Imitating countries cannot retain the old technology which is actually inferior, and they have little chance to compete internationally with the new technology as long as they are not able to introduce contingent technologies which make the new general purpose technology more appropriate to their own factors endowment, or alternatively, change their system of relative prices.

In these circumstances technological variety shrinks in international markets and with it the scope for advantages from trade in imitating countries. Innovating countries on the other hand, able to command the new technology and to extract the maximum levels of total factor productivity levels out of it, have a huge interest in higher levels of globalization. The larger the flows of trade in international markets and the greater the exposure of domestic markets to international competition, clearly the larger are the opportunities for increasing market shares and profitability of innovating firms, based in countries where the relative prices of production factors fit better the specific mix of output elasticities of the new technology.

Total factor productivity levels of a given non-neutral technology differ among countries according to the differences in relative prices. The closer the structure of the economic system to that of the innovating country, the smaller are the differences in total factor productivity levels. For countries characterized by a significant difference in relative prices, the new technology yields lower efficiency levels which must be compensated for by the absolute levels of factors prices. Competition in international markets is clearly affected by such differences in relative prices and hence actual total factor productivity levels. The larger the differences among countries in terms of factors prices, the larger are the asymmetries in the gains from international trade.

The effects of contingent technologies

When contingent technologies are being introduced, non-innovators cannot withstand the threat of the competitive advantage of innovators. Here the advantages of trade can be sharply asymmetric and only innovators have an actual advantage in increasing the flows of international trade. In such conditions, because imitation is not profitable, non-innovators can cope with international competition only if they can innovate in turn, either with the introduction of shift technologies, locally neutral, or with other contingent technologies better able to extract all the productive capabilities of their own structure of relative prices.

When innovating countries are able to generate contingent technological changes they are likely to better retain their competitive advantage which consists in a better combination of their specific economic structure and hence relative prices and the specific characteristics of the technology. Contingent technological changes cannot be imitated. Contingent technological changes cannot be imitated simply because prospective adopters cannot take any advantage of such a technology which is not appropriate to local factors endowments. From this viewpoint contingent technological change yields substantial barriers to imitation which in turn becomes a pervasive factor of barriers to entry in the new market for firms based in countries with a completely different price structure.

The distinction between contingent and general technological change makes it possible to highlight a major difference in the role of innovation and imitation. Imitation is sufficient for followers to reduce asymmetries in international markets only after the introduction of a perfectly neutral general purpose technology: that is, a new general technology which is neutral with respect to all the techniques in place before its introduction. This in turn implies that adopting countries already had a structure of endowments and relative process of production factors close to that of innovating countries.

In this case innovators can retain only a transient competitive advantage which lasts until all potential adopters are able to use the technology. It is clear that the faster the adoption, the higher are the diffusion rates and the shorter the time spell of duration of the asymmetric distribution of the gains from trade in favor of innovators.

When technological change consists both of shift and bias effects, at least for adopting countries, followers face emerging asymmetries which last as long as not only they are able to imitate, but also to introduce incremental contingent innovations which adapt the general purpose technology to the local factors markets and/or reduce the differences in the relative prices with respect to innovating countries.

Finally, imitation cannot help followers when contingent technologies are being introduced in innovating countries. When technological change is contingent laggards can only face emerging market asymmetries with the introduction of other innovations. In such circumstances international trade can be considered beneficial at large for all countries only in the long term if and when technological change can be considered endogenous and putty.

The conditions for symmetric advantages from trade

A general statement can be made now: the introduction of new general purpose technologies which reflect the local endowments of innovators can put at risk the comparative advantages of countries with different endowments and different, albeit obsolete and inferior, technologies. Traditional assumptions about the variety of technologies, consistent with the variety of endowments and hence relative factors costs, are less and less valid at the time of the sweeping introduction and diffusion of new general and highly productive technologies which are characterized by strong factors intensities. Technological variety risks collapsing in these circumstances, to a single general technology with high levels of total factor productivity. Only countries and regions within countries where most productive factors are relatively cheaper can take full advantage of such opportunities. Countries where most productive factors are relatively more expensive, that is regions which are becoming peripheral, can face a decline in their relative efficiency. Only if such countries losing ground in international markets, after the introduction of a new radical and yet highly biased technology, can react and innovate in turn, can international trade yield benefits for peripheral countries.

More specifically two conditions emerge: (1) domestic firms must be able to change their technology and (2) local factors prices can be changed. Technological change can be considered, at least to some extent, the endogenous outcome of the innovation capacity of the system as shaped by the interplay in the market-places of the interactions of firms that are able to change their technology. The structure of relative prices instead is very much beyond the direct influence of the spontaneous interactions of the firms in the market place, at least in the short term. This is the scope of economic policy.

When and if technological change is endogenous, and domestic firms are able to actually introduce new, better technologies more appropriate to their own factors endowment, from a dynamic viewpoint it seems clear that any increase in the actual levels of international competition – made possible by the reduction of barriers, customs, transportation and communication costs and in general by better conditions for the circulation of goods and services – is likely to have important effects in terms of rates and direction of technological change. The increase in international competition favors countries which specialize in technologies that make larger use of locally more abundant production factors. Competitors facing the shrinking of their market shares will try and introduce new technologies.

The direction of the new technologies is likely to be biased in favor of a most intensive use of locally abundant production factors. This bias is the result of two arguments. First, firms are induced by the reduction of their market shares to introduce new technologies. Second, firms which happen to introduce technologies that make a more productive use of production factor more locally abundant will experience a competitive advantage with respect to firms that do introduce new technologies, but with the "wrong" bias. For a given direction it is clear that firms able to introduce new technologies with a larger potential and actual total factor productivity growth will win in the competitive race.

Increased technological specialization is the clear outcome of a race where each economic system moves along a technological path shaped by the characteristics of the endowment of the production factors. A direct relationship between the reduction in barriers to trade and the rate of technological advance can also be identified, provided each economic systems is able to react with the introduction of new technologies to the effects of increased competition in international markets.

Selective flows of international trade can be beneficial at large when they concern specific intermediary inputs which are associated with the introduction of new technologies in downstream industries. In this case the selective disclosure of domestic markets can yield positive effects on the domestic economy when the indirect effects of the decline in the market prices of the new key intermediary inputs are likely to yield significant benefits in terms of the actual increase of the total factor productivity levels, as induced by sheer price effects, in downstream sectors.

An important result of the analysis here consists in highlighting the conditions for international trade to yield actual benefits for all parties engaged when new general technologies are being introduced. At the time of the introduction of a new general purpose technology which is shaped by the specific factor intensity of a idiosyncratic country and yet has such shift effects to yield larger output in all countries of application with all possible relative prices, international trade can have dramatic effects on the global economy.

Technological variety is swept away by the sheer superiority of the new technology. The opening of domestic markets to international competition is likely to yield a major asymmetry in the distribution of the gains from international trade. Production in the innovating country is clearly more efficient and the domestic markets of adopters are exposed to strong international competition by innovating firms.

Imitating countries have an incentive to accept international competition only if and when local firms are actually able to minimize diffusion lags and to change the new technology, possibly with the introduction of contingent technological changes which reshape the mix of output elasticity of production inputs in a way that fits better the local factors endowments. Innovation policies are a necessary condition for imitating countries to enable to participate in international trade. The second condition is clearly an economic policy aimed at changing the local relative prices towards levels which are closer to the original specification of the new general purpose technology. Finally, selective strategies of participation in international trade, for intermediary inputs that are key factors in new general purpose technologies, can yield important effects in downstream industries, even if they expose the country to the overwhelming pressure of exports from innovating countries in those specific products markets.

The situation is completely different when contingent or locally neutral technologies are introduced. A clear divergence of interests emerges instead between countries able to generate locally neutral and/or contingent technologies and competitors with respect to international trade. Contingent technologies consist of a change in the output elasticity of production inputs. Contingent technological changes, as already considered, do yield an actual increase in total factor productivity and hence do lead to lower production costs, but only when the structure of relative prices is consistent with the new shape of the production function. Contingent technological changes can benefit only countries with quite specific factors endowments and do not apply to all possible factors markets.

Conditional convergence

The differences in the effects on international competition between contingent and general technological changes yield important results in the context of the convergence debate. This interesting debate has followed the results of the empirical investigation carried out by of Baumol in 1986 on convergence in the rates of growth across countries. Baumol put forward the hypothesis that technological change during the nineteenth and twentieth centuries offered important opportunities for less advanced countries to reduce their lags, with respect to rich countries, in terms of labor productivity and general per capita income. The convergence hypothesis was then formulated and the evidence of the negative relationship between the levels of labor productivity and their rates of growth was provided for a number of selected and yet significant countries. Conflicting evidence was then provided by the findings of DeLong (1988) and Friedman (1992).

Ruttan (2001) reviews the main results of the debate and suggests that the actual convergence may be conditional upon the rates of savings and investment in both human and fixed capital. The notion of conditional convergence seems most useful when the specific characteristics of technological change are taken into account. The basic claim here is that convergence is conditional to the extent to which technological change mainly consists of shift or bias effects. When a new radical technology is introduced with a mix of output elasticity which fit the specific conditions of the innovating countries a strong push towards a short-term divergence is likely to take place. Alternatively, when technological change mainly consists of the widespread introduction of contingent technologies, a structural trend towards divergence is likely to be set in motion.

The introduction of new general purpose technologies favors the countries which have "appropriate" endowments and as such are better able to extract all the potential productivity growth from the new technology. Competing countries and regions can try and contrast this process with two strategies.

A strategy aimed at increasing the flexibility of relative prices is clearly the first. Competing countries can try and imitate the relative prices that match the new bias in the technology. This strategy is viable when the relative endowments of the local economy are under the control of the tools of the economic and fiscal policy. When the local endowment of skills is too low and hence the relative wages for skilled labor is too high, an active training and fiscal policy can help the change in the relative prices. This is also the case when strategic inputs are supplied in monopolistic conditions and upstream producers enjoy substantial barriers to entry and are not able to adopt better technologies and or to reduce their costs inefficiency. A selective competition policy can help reduce such relative prices, together with an innovation and diffusion policy aimed at increasing the rate of technological change, either by means of local innovative efforts or the accelerated adoption of new existing technologies, targeted in upstream sectors which supply the rest of the economy of such key intermediary inputs. Imports of such intermediary products from more efficient countries can also be very efficient in this context

Competing countries, however, can imitate the new general technology and can cope with it if contingent technologies are introduced so that the local endowments of production inputs can cope with the characteristics of the new general purpose technology. The divergence is likely to last only for as long as is necessary for followers to imitate and introduce in turn contingent technologies which reduce the relative price inefficiency brought about by the introduction of the new technology. Second, and more important, followers can find a competitive advantage in the differences in absolute costs of production factors. Even though the new general purpose technology fits better in different structures of relative prices, the imitating and follower country can compensate with the sheer levels of absolute factors costs for the competitive advantage of the innovators.

When instead leading countries are able to introduce locally neutral and/or contingent technologies they can retain a long-lasting competitive advantage. Here divergence can prevail because followers can only revert to their own technology as a way to contrast the new competition. Followers here have no chance to imitate and innovate, taking advantage of learning to use the new technology, after imitation. Divergence can take place with widening gaps among countries in terms of per capita income and labor productivity. The risks of falling away from the convergence club in this case are clearly stronger the larger the difference is in the relative prices among countries.

The notion of conditional convergence, which depends not only on the rate of technological change, as in the neoSchumpeterian tradition of analysis developed by the economics of innovation, but more specifically, on the direction of the new technologies being introduced, seems to provide a new set of hypothesis for empirical investigation.

Globalization and regional integration such as in the case of the European Union can generate positive results only if and when a number of risks and possible shortcomings are well understood and properly balanced with appropriate policies. Globalization is likely to increase the exposure of each economic system to the competitive advantage of innovating countries on two counts. Innovators can take advantage not only of lead times and imitation lags of competitors, but also of the specific cost advantage provided by the direction of the new technologies being introduced, when they make intensive use of locally abundant and hence less expensive production factors. On a closer look it seems also clear that the lead times of innovations and the direction of the new technology can reinforce each other: innovators have a longer time span to make the most productive factors more abundant and hence less expensive.

Countries and regions can take advantage of globalization and regional integration, in terms of higher levels of specialization and larger niche markets for their products, only if they are ready to face the competitive lead of innovating countries with the acceleration of their own rate of introduction of new technologies in a well-defined direction, in terms of output elasticity, one which impinges upon their specific factors endowment and is able to make the best use of it.

By way of conclusion an hypothesis can be put forward: if a sequence is likely to take place between the introduction of new general purpose technologies and the subsequent array of contingent technological changes, a sequence between periods of divergence and periods of convergence can be found in historic time. The introduction of new general technologies, locally neutral but with high levels of total factor productivity will favor innovating countries where the most productive inputs are also cheaper. Composition effects favor innovating countries together with epidemic diffusion lags. Imitating countries are unable to match the competitive advantage of innovators, except when absolute factors costs are so low as to compensate for the effects of technological change both in terms of rate and direction. A period of divergence in terms of rates of growth among countries and regions is likely to take place at this time. Innovating countries are able to forge ahead and imitating countries lag behind. The eventual introduction of contingent technological changes in imitating countries, however, can reverse the process, together with the efforts to close the gap in terms of the difference of the relative prices of production factors. At this time laggards may be able to reduce the collapse of technological variety and take advantage of positive composition effects. Market shares increase labor productivity and profitability can be improved. A period of convergence can follow the first phase of divergence and laggards can reduce the gaps with respect to the countries where the general purpose technology had been originally introduced.

Conclusions

Comparative advantages make it possible for heterogeneous economic systems to interact and trade with symmetric advantages. Comparative advantages, however, are possible only when and if heterogeneous countries are able to implement their own technology. The basic assumption for trade to be symmetric and mutually beneficial is that each region is able to use a technology which reflects the local endowments of production factors and hence the relative levels of production factors. Absolute factors prices can compensate, to a limited extent, the lack of appropriate levels of technological heterogeneity. In this context it seems clear that the introduction of new general purpose technologies which reflect the endowments of innovators and yet have high levels of total factor productivity can yield negative consequences.

The adoption of the new technologies is necessary and yet it is not sufficient to restore the basis of local competitive advantages and only the sharp reduction of the absolute levels of factors costs might compensate for the new unbalance. The introduction of contingent technologies, together with the adoption of the new general purpose technology, is the single strategy for adopting countries to re-establish the basis for comparative advantages.

This analysis has important implications for intellectual property rights. In a global economy where substantial barriers to imitation and to entry arise from differences in relative factors prices and biased technological changes, strong international protection may impede and delay the necessary efforts of adaptation of the new technological knowledge to the local context. The recombination of external knowledge with local, domestic knowledge is a necessary condition for the introduction of a technology biased in favor of the locally abundant production factors. The introduction of contingent technological change by followers might be barred by too strong a protection of the patents held by innovators on the general purpose technologies they contributed to introducing. Only if strong international protection of intellectual property rights is associated with a strong incentive to trade the technology and make the recombination possible, may a strong regime for intellectual property rights become beneficial.

10 Feedbacks, path-dependence and evolution

The full circle of feedbacks and interdependent causation between technological change and structural change can now be articulated. Building upon the localized theory of technological change elaborated in previous contributions (Antonelli 1995, 1999a, 2001a) we assume that unexpected changes and disequilibrium conditions in products and factors markets act as the basic inducement mechanism that pushes myopic firms to innovate and hence to the generation, introduction, imitation, adoption and adaptation of new technologies.

All changes in the structure of the economic system provide the basic inducement to the introduction of innovation. The rate of introduction of new technologies is directly influenced by the amount of changes, with respect to their myopic expectations, firms, constrained by substantial irreversibility of their tangible and intangible capital stocks and by their employment levels, face in their factors and products markets. In turn at each point in time the structure of the economic system is a prime causal determinant of the direction of technological change. Depending on the specific characteristics of each economic system in terms of endowments of basic inputs, industrial structure and market structures for intermediary inputs, capital goods and skills, firms will select the type of technological change which is more profitable. Firms will generate, introduce and adopt the technologies which fit better in each given structural context.

The model elaborated in this book provides a synthesis of the notions of internal and external path-dependence. Internal path-dependence takes place when the path along which the firm acts and eventually innovates is determined by localized learning and the irreversibility of its production factors. External path-dependence is instead determined by the external conditions which dictate and shape the successful introduction of new technologies at the system level. The model discussed here works both on internal and external path-dependence. Internal path-dependence is appreciated because of the role of irreversibility, localized learning and switching costs that are specific and internal to each firm in inducing the rate of introduction of new technologies. External path-dependence here is determined by the role of factors endowments and relative prices that induce the direction of technological change.

The new global economy provides a context in which an international industrial dynamics approach seems necessary. Firms react to all changes to their myopic expectations not only with changes in the price-output mix but also with the introduction of new technologies. The new technologies reflect the specific context of action. Such a context is characterized by competing firms active in broad and complex industrial structures which take into account a variety of regions and countries with significant heterogeneity in both technologies and endowments.

The introduction and adoption of a new technology has in turn relevant effects in both products and factors markets. The introduction of new technologies affects both the absolute and the relative prices of important production factors and hence the competitive advantage of downstream customers. This is true in general and especially when it concerns key industries, which provide the rest of the system with important intermediary inputs embodying new technologies.

The rates of entry and exit of firms into the new industries and from the old ones is an important structural and dynamic characteristic of the system which affects its overall performances. In general the price flexibility of all inputs is relevant in this process.

By the same token the introduction of new technologies has important effects on the markets for basic inputs and especially for skills. The shape of the long-term supply schedule of skilled and trained labor plays a key role in assessing the overall effects of the introduction of new technologies. Systems where the supply of labor is price anelastic and does not adjust in the long term to the shifts in the derived demand for skilled labor engendered by the introduction of skill-intensive technologies have clearly far different and worse performances than skill elastic systems, at least in terms of rates of diffusion of new general technologies. Only the introduction of new technologies can help these countries to face the new competitive arena.

The introduction of new technologies affects the single system in a variety of ways that are defined by such parameters as the time elasticity of entry and exit of firms and the long-term elasticity of supply of skills. These changes will in turn modify the relative prices and expose firms to new sources of disequilibrium and variety. The variety of players and of their reactions in the structure of the system will induce firms to cope with emerging disequilibrium conditions by means of the introduction of new technologies. The rate of the introduction of innovations is induced by the levels of discrepancy between expectations and irreversible commitments while the direction is shaped by the specific composition effects at play at each point in time.

Horizontal effects concern the industrial dynamics among firms which sell similar products: the introduction of a new technology within each industry pushes all the other firms to change their technology in turn, either by means of innovation or by means of imitation. The adaptive imitation is especially important when composition effects matter. Horizontal effects are enhanced and magnified in the global economy. Firms located in a variety of factors markets compete in single international product markets. Here the dynamics of structural and technological change includes the diffusion of innovation. In the global economy the structure of each system is exposed to international competition where the interaction between the rate and the direction of endogenous technological change and the composition effects are most relevant.

Vertical effects affect both direct customers and indirect ones in downstream industries, in terms of relative and absolute factors costs, as well as direct and indirect suppliers in upstream industries. The whole input-output matrix can be affected by the introduction of a new technology with snowball effects.

Firms can build transient competitive advantage when the imitation of new general technologies is delayed by information asymmetries. In a complex economic system, where heterogeneous factors markets matter, contingent technologies, instead, provide long-lasting competitive advantages to innovators because imitation and entry are barred by composition effects and hence only probit diffusion can take place.

Advanced core countries can now be identified by their innovation capabilities defined in terms of the levels of coherence between the specific characters of the technology being generated and introduced at each point in time and the features of the economic system where the innovations are being introduced. Core countries better able to master the co-evolution of technological change and structural change can build up a long-lasting competitive advantage which combines the composition effects with the bias of the new technologies and the shift component of each innovation.

Imitating countries are obliged to follow up and can compete only if and when they are able to converge towards both the structure of relative prices of leaders and to adopt the technologies and adapt them to their own structural characteristics. Convergence to the labor productivity levels of advanced countries is the joint outcome of both technological and structural adaptation.

The global economy provides a conducive context for such path-dependent interactions for the variety of factors markets and technologies which confront each other on single international product markets. The larger the variety of factors markets and technologies, the stronger is the inducement mechanism for the introduction of new technologies and the greater are the incentives to master the co-evolution of economic and industrial structures and of technology.

The basic foundations of a recursive and path-dependent process of localized interdependence and causation between the introduction of new technologies and their adoption and the structural and industrial dynamics of economic systems are now set. Composition effects, the rate of introduction of innovations and the direction of technological change are the three key elements of a continual and self-propelling process of change.

At each point in time the changes of the structure of the system push the firms towards the introduction of new technologies. The characteristics of the system induce the choice of the direction of the new technologies. The new technologies may eventually change the very basic structure of the system itself. Structural change in turn feeds technological change. The system is never able to attain an actual steady equilibrium. The continual dynamics of path-dependent feedbacks between structural change and technological change pushes the system along a path which is characterized by the initial condition of the process and by the specific characteristics of the changes introduced at each point in time.[1]

The merging of the economics of innovation with the economics of technical change delivers an important result. It makes it possible to grasp the path-dependent and recursive process of dynamic interdependence and causation between structural change and technological change. Economics of innovation contributes the endogenous understanding of the processes that lead to the generation, introduction, adoption, diffusion and adaptation of new technologies. Economics of technical change provides a unique analysis of the key role of the economic and industrial structure in assessing the effects of the introduction of new technologies. In the broader context of analysis provided in this chapter the structure of an economic system is clearly at the same time the determinant and the effect of endogenous technological change.

The analytical tools elaborated in the microeconomic tradition to implement the representation of the economy as a closed system can be manipulated and recombined with an approach which takes into account the capability of agents to generate endogenous technological changes that reflect their own specific conditions so as to lead to a process approach which presents the working of the economy as an open system.

Part III

Applications and implications

11 Understanding the economics of new information and communication technology in the global economy

The digital revolution is engendering a global digital divide. It is more and more evident that the introduction of new information and communication technology in the global economy parallels widening and ever increasing asymmetries among countries and even regions within countries.

The comparative economics of new information and communication technology in the global economy can provide much empirical evidence to appreciate the relevance of the analytical tools so far introduced. A comparative approach to assessing the gaps among countries in the rates of introduction of the new technology, in the ability to contribute to the construction of the new technological system on the one hand and in the capability to take advantage of the new technology, seems more necessary. The international analysis of the effects of the introduction of the new technology, both in terms of diffusion, adoption and impact on total factor productivity growth should easily confirm relevant gaps and asymmetries among countries. Such a new gap is worth a detailed analysis which the tools discussed in this work may help.

New information and communication technologies can be portrayed as a locally neutral general purpose technology which consists mainly of a radical shift in the production functions of a large array of goods and a minor bias, at least for a few advanced countries. New information and communication technology in fact can be stylized as a non-neutral, capital and skilled labor-intensive technology for most countries. Clearly it is a technology which exhibits significant bias effects when applied outside of the original context of first introduction and implementation, the US (David 2001; Jorgenson 2001; Quah 2001).

New information and communication technology fits well into the definition of general purpose technology from three viewpoints. First it is the result of an array of complementary and interdependent technological innovations which have gradually converged towards a new technological system. Second, the pervasivity and fungibility of new information and communication technology is very high: almost all economic activities are influenced by the applications of this general technology. It consists of both product and process innovations where many product innovations in upstream markets engender process innovations in downstream industries. Consequently it applies to a large variety of products and processes with high levels of profitability of adoption. Finally, new information and communication technology is inherently global: its introduction engenders a systematic and relevant growth of total factor productivity and reduction in costs that its adoption is profitable in a great variety of regions. New information and communication technology is a superior technology in a great array of products and factors markets.

The extent to which firms active in different factors markets and in different products markets can take advantage of it, however, differ. New information technology, because it is a general purpose technology, is likely to engender major asymmetries and increasing variance in the economic performances of trading partners in the global economy.

Some firms, located in some regions and active in some industries, may be better off than others. The direction of this technology, in terms of the new bias in the use of production factors and the context of adoption, in terms of the relative costs of the different production factors, play an influential role in shaping the ultimate effects of the adoption of new information and communication technology on the actual range of different levels of competitive advantage and of the general efficiency of adopting firms.

The introduction of new information and communication technologies is likely to have significant effects in terms of a sharp reduction of technological variety. As such, new information and communication technologies can affect the source and the structure of the comparative advantages of regions where many specific productive inputs, associated with the new technology, are more expensive at least in relative terms. From this viewpoint the introduction of new information and communication technologies in the global economy, because of their absolute superiority and despite their bias, can engender new relevant asymmetries among regions, not only with respect to the pace of diffusion, but also and mainly in terms of the capability of each region to extract appropriate economic benefits from their necessary adoption. The evidence on the digital divide, with the ever increasing gap between digital-rich and digital-poor, finds here some basic explanation.

New communication technologies can be characterized as a new technological system which is the result of a sequential stream of introduction of new complementary technologies aligned along a technological path shaped by similar characteristics in terms of the prevailing output elasticity of production factors. New information and communication technologies are the result of a complex and systemic process of implementation and continual elaboration which is strongly embedded in the fabric of economic, institutional and social systems of interactions which reflect strongly the localized characters of the innovating countries. They can be considered a clear example of a new technological system, characterized by the complementarity and interdependence between an array of new technologies and specific fields of technological knowledge, which has emerged in a context shaped by strong idiosyncratic features. As such the bias of the production function necessarily reflects the endowments and the process of the innovating countries (David 1987; Antonelli 1993a, 1993b, 2001).

Economics of innovation has analyzed in depth the process of generation of the new technological system and has provided a vivid collection of outstanding case-studies (Fransman 1995 and 1999). The analysis of knowledge as a collective good has been applied successfully in this context (Abbate 1999), as well as in the understanding of the role of standards (Grindley 1995). The cumulative character of technological knowledge and the role of learning processes has been documented and confirmed (Wasserman 1985; Mowery 1996; Cusumano and Selby 1995).

The role of geographic proximity as an important factor conducive to localized increasing returns in the generation of the new technology has received considerable attention (Swann *et al.* 1998). The interplay between technological complementarities and technological convergences and the strategies of firms, exposed both to continual cross-entry from adjacent markets and adjacent technologies, has provided important insight both on the understanding of the generation of the new technological system and on the theory of the firm and the markets exposed to rapid technological change (Shapiro and Varian 1999).

The co-evolution of industrial structures and the technology has been documented in many cases (Malerba 1985; Dorfman 1987; Mowery 1996). The sequence of monopolistic market structures followed by differentiated oligopolies and eventually monopolistic competition across product niches has been confirmed and there is strong evidence indicating the key role of entry and imitation (Klepper and Graddy 1990).

Entry has also been made possible by an intense flow of product innovations targeting well-selected market niches where groups of heterogeneous consumers gather. Niche-pricing has proliferated and a variety of profit rates can be found across single products markets which are only partly substitutes.

The reverse cycle has also been confirmed, however, and the sequence of a large variety of rival innovations, eventually reduced by intense selection processes where by means of the replicator dynamics a few leading firms have been able to consolidate a long lasting market power, has found relevant applications (Crandall and Flamm 1989). The role of technological selection among rival and competing technologies has been investigated and assessed and the relevance of the processes leading to the emergence of a dominant design and related dominant clubs of firms have been confirmed in this field (Mansell 1994; Duysters 1996).

The analyses of the characteristics and the determinants of the diffusion of new information and communication technologies, across regions and countries, industries and firms, have proved their effective use (Antonelli 1990 and 1991). The study of information and communication technologies has also made it possible to investigate the limits of traditional demand analysis and to appreciate the role of interdependence in utility and hence in consumption patterns, leading to the notion of network externalities (Antonelli 1992).

The effects of the introduction of new information technologies at the company level, on the organization of corporations, including internal and external labor markets, on the production process and on relations in the intermediary and final markets between suppliers and customers, including the effects on growth strategies, have been the object of a variety of analyses. These effects have been studied with care, as have also the characteristics of the accumulation of new knowledge (Antonelli 1988; Allen and Scott Morton 1994, Bresnahan *et al.* 1999).

The effects of the introduction of new information and communication technologies on the organization and architecture of the production process are most important. Lean production is generalized with significant reductions in the stocks of both final and intermediary products. Customized mass production is applied to a variety of goods and parallels the generalized adoption of multiniche strategies in global markets. Modular manufacturing is systematically implemented and the architecture of the manufacturing process is redesigned according to the structure of modules. Interactions between the marketing units, manufacturing and research and development laboratories are closer, and the feedbacks are tighter so that times to market for new products shrink radically.

After the introduction of new information and communication technologies the best practice in the production process in fact is more and more characterized by higher levels of vertical division of labor and specialization with the pervasive use of highly qualified knowledge-intensive business services. This trend is especially relevant in the new service industries where skilled labor intensity is very high and exhibits well-defined features in terms of formal training and academic background. Next, and most important, the new service sectors and especially the new knowledge-intensive-business service industries are global in character, with a direct access to international markets, a foot-loose location of different branches and specialized units across countries and rapid rates of exit and entry in local markets according to their endowments of an array of intermediary inputs and their relative prices. The increasing role of such intermediary inputs makes the composition effects even more important from an industrial organization viewpoint. New information and communication technologies are characterized by the pervasive use of advanced communications and information-processing services as key intermediary inputs.

Systematic applications of new information and communication technologies to monitor and implement market transactions make it possible to shrink search and comparison costs and to increase the scope for the absolute and comparative quality assessment of the products available in the market place. The development of electronic markets makes it possible to implement multilateral exchanges with significant effects in terms of the number of players in the market-place both on the demand and the supply side. As a result intermediary markets become more reliable. Bilateral relations moreover can be better implemented by means of on-line links and the terms of the contracts can be monitored in real time with further reductions in information asymmetries between contractors and between *ex ante* and *ex post* contractual conditions. Transaction costs decline and production processes can be better specialized.

Systematic applications of new information and communication technologies within corporations make it possible to reduce information asymmetries in principal–agent relations, with substantial reductions in agency costs and increased transparency of internal labor markets. Flat organizations substitute multi-layer hierarchical structures. At the same time existing hierarchical structures can extend their reach and command larger and more complex coordination processes.

Multinational companies, mainly based in the US and UK, rely upon new information and communication technologies to organize their activity globally. distributed in a variety of quasi-decentralized manufacturing and service units nested into a network of international flows of intermediary and final products, manufactured in selected countries according to the local systems of relative prices.

Mass communication makes it possible to globalize the brands and reputation of a few global companies that sell worldwide customized varieties of a limited number of products based upon the recombination of homogeneous modules. The concentration of brands parallels the concentration of mass media with a sharp reduction in the number of players in the final markets. The global integration of distribution chains, marketing services, advertising and broadcasting makes it possible to take advantage of huge economies of density in the global markets for final products. The wider the scope of application of a brand, both in terms of regions and products, the lower are the unit costs and hence the more effective and dedicated can be the investments and the higher are the competitive advantages of global players on local competitors.

Finally, the introduction of new information and communication technologies has important effects on the new emerging markets for technological knowledge. Tradability of technological knowledge is improved by means of on-line inter-actions, as well as appropriability. Reductions in transaction costs and thicker markets for technological knowledge increase the scope for specialization and hence efficiency in the production of knowledge itself with significant cumulative effects in terms of rates of introduction of new technologies.

Important empirical evidence has been gathered about the economic effects of new information and communication technology, mainly if not exclusively in the US economy (Brynjolfsson and Yang 1996; Attewell 1994; Marchesi 2002).

At the aggregate level much empirical work has been conducted in the US, in order to assess the effects of new information and communication technology on total factor productivity growth (Jorgenson 2001). The empirical evidence suggests that the growth of the information technology sector has been especially fast, albeit in a context marked by the slowdown of other important sectors of the economy (David 2001). The evidence has been especially strong on the decline of monetary and especially hedonic prices for products and services delivered by the information and communication sector to the rest of the economy, both to the household and to the other intermediary users. Evidence about the general effect of new information and communication technology since the mid-1990s, however, does confirm that the new technology has played a key role in the resurgence of the US economy (Jorgenson and Stiroh 1995 and 2000).

At the firm level the introduction of new information and communication technologies has direct effects on the productivity levels. The variation among firms in their ability to take advantage of the new technology on the users' side has been specifically investigated and confirmed (Lichtenberg 1995; Brynjolfsson and Hitt 1995). The empirical evidence gathered in France confirms the positive effects on total factor productivity growth (Greenan and Mairesse 2000). Differences among firms, however, are relevant and their characteristics do play a specific role in enhancing the effects of new information technologies. At the firm level the asymmetric effects of the new technology, in terms of general efficiency, are well identified.

According to the main results, the introduction of the new technology parallels an increased segmentation in the labor markets with higher demand levels for skilled labor and a rapid shift away from unskilled labor. The segmentation of the labor markets in turns coincides with an increased variance in wage levels. Wages for skilled labor have increased far above the average rates of increase in wage levels and some reductions in wages for unskilled labor, in real terms, have been noticed. An intense debate has been taking place about the specific relevance for new information and communication technologies of the "skill-biased technological change" hypothesis. The hypothesis of a complementarity between new technologies, fixed capital and skills had been put forward first by Griliches (1969) and applied to analyzing the effects of the introduction of the dynamo by Goldin and Katz (1996 and 1998). The skill-biased-technological-change hypothesis has been further elaborated by Caselli (1999) who suggests that the characteristics of the technology matter in assessing the skill bias hypothesis. Caselli (1999) argues that the characteristics of the technology do matter in assessing the changes in the labor markets that have been taking place in the US economy in the 1990s. Acemoglu (1998) elaborates the hypothesis that the skill-biased-technological-change is not exogenous, but on the opposite the intentional result of the introduction of innovations which make use of the most abundant local factor: this is the case with academic skills in the US economy. Along these lines many empirical studies (Krueger 1993; Autor *et al.* 1998) show that the skill bias takes place because the new technology requires high levels of formal learning based upon academic training. Because of the requirements of the new information and communication technologies, on-the-job training and learning by doing and by using can take place effectively only when appropriate levels of human capital have been accumulated. The new characteristics of the production process have powerful effects on the demand for labor in terms of an increasing polarization: a fall in the demand for labor with low levels of formal training and the steep increase in the demand for labor with high levels of formal skills. Bartel and Lichtenberg (1987 and 1990) stress in the same context the role of the novelty or age of the technology. In the early stages of introduction new technologies require higher levels of skills because of the risks and complexities associated with the necessary learning processes. Eventually however the skill bias should decline as well as the premium associated with the skill contents.

The evidence on the evolution of labor markets and wages confirms the hypothesis that the introduction of new information and communication technologies acts as a powerful factor in increasing segmentation and asymmetries of the labor markets within the economy of advanced countries, well ahead in the process of introduction and adoption.

The evidence for the strong relationship between rates of introduction of the new technologies and increased variance both in wages and productivity levels within the economy of advanced countries is now large and convincing. Little analysis has been devoted to the international dimension of the very same dynamics. The empirical analyses on the effects of new information and communication technologies, however, have paid very little attention to the comparative perspective and assessing the relative effects of new information and communication technologies in the US economy with respect to the rest of the world (Wyckoff 1995; Ruttan 2001).

In the global economy the asymmetric and divergent consequences of the introduction of new general purpose technologies, such as new information and communication technologies, are stronger because of two intertwined dynamics: (1) the delay in adoption due to sheer information asymmetries and differentiated profitability, partly explained by (2) the heterogeneity of endowments for primary production factors and of industrial structures for intermediary inputs and hence of relative factors prices which engender the powerful dynamics of composition effects. In the rest of the world, in fact, and especially in the developing countries, the diffusion of new information and communication technologies is lagging behind. In countries where the telecommunications network is still missing, the resources for the basic infrastructure and the specialized skills are scarce. The digital divide is emerging with increasing effects in terms of total factor productivity differentials and average production costs and hence diminishing opportunities for growth in the global economy. The need to push the necessary recombination of external and domestic technological knowledge in order to apply the general knowledge available to the local structure of endowments and the subsequent introduction of contingent technologies is stronger and stronger (Mansell and Wehen 1998).

Here the integration of the economics of innovation and of the economics of technical change can play an important role in elucidating the systemic characteristics of the new technology in the global economy. The study of the determinants of the introduction of such a new radical technological system in the global economy can help in grasping the strong asymmetries between the virtuous dynamics of innovation, fed by the accumulation, generation and distribution of new technological knowledge, mainly in the US economy and the slow pace of advance in the European economy and especially in the rest of the world.

The study of the effects of the introduction of such a new radical technology in the global economy provides a clue to understanding the key role of the structure of each economic system into which the new technology is applied. Countries differ widely in terms of their ability to take advantage of the new general purpose technology and many are lagging behind while the US economy is forging ahead with a new resurgent vitality.

A framework, in a comparative context, is necessary which is able to try and explain why countries differ both in the capability to contribute to the rate and direction of technological change and in the capability to take advantage of the new technologies. Here the distinction between contingent and general technological change matters. General technological change takes place when all techniques of the new technology are superior to the previous ones. This is the case of the US economy, where the new technological system has been first introduced and implemented, reflecting the local endowment of production factors. The US economy has had the opportunity to elaborate a new superior technology which makes a more intensive use of the specific factors endowments of that economic structure.

The effects of lead times here are most relevant in a global and comparative context. Not only can the US economy take advantage of the early command of the new technology and the imitation lags of the follower. Lead times have also provided an opportunity to adjust the economic structure to the specific requirements of the technology, in terms of relative output elasticity, of an array of production factors. The unique features of new information technology which have emerged in such a process, in terms of relevant shift effects, are such that the adoption of the new technology is profitable in most countries and industries. The strong bias effects, built into the technology, for prospective adopters located in a different economic context, however, are such that adoption does take place and yet it yields lower levels of actual total factor productivity.

The introduction of a new radical and general purpose pervasive technology, able to activate a widespread convergence of an array of different technologies towards a single dominant design, is at the same time the cause of a general opportunity for growth in the global economy but also the cause of widening gaps and asymmetries across countries, according to the difference of their relative factors prices with respect to the structure of the innovating economy. The diffusion of such a pervasive and superior technology reduces the technological variety upon which the competitive position of countries with different economic structures and relative factors prices could rest.

Economics of innovation and technological change should pay attention to the analysis of new and information technology from the viewpoint of the specific combination of production factors at play and their relative output elasticity. New information and communication technology can be considered as a new technology which pushes a variety of different production functions towards the use of techniques that are relatively more capital-intensive and especially highly skilled-labor-intensive. On the contrary, the application of new information and communication technology tends to reduce the need for low-skilled labor. More specifically new information and communication technology reduces the scope of usage of labor with low levels of education and formal training. Here there is a further important distinction between human capital, whether its accumulation consists mainly of on-the-job learning processes with low levels of formal training in University and higher education institutions, or human capital with higher levels of basic education. New information and communication technology-based innovations make highly intensive use of very specific and dedicated skills with high levels of requirements of formal academic training in well-defined disciplinary fields. The command of tacit knowledge, without proper academic background, is difficult and loses most of its relevance in this field of application, although it had a major role in traditional engineering technologies (Antonelli 1988, 1991 and 1992).

Next, and most important, new information and communication technology, as a general purpose technology, is highly pervasive and applies to a variety of different production functions. The introduction and adoption of new information and communication technologies requires specific innovations which consist in the elaboration of specific applications and in the usage of new intermediary inputs and new dedicated capital goods.

The strong capital-intensity of new information and communication technology cannot be considered only relevant at the abstract levels. On the contrary the specific goods which constitute the new relevant basket of capital is highly specific and consists mainly of a large, capillary and extended network of telecommunications infrastructure which connects almost all the relevant production and consumption sites.

Specifically we see that the successful introduction of new information technology is the result of the careful blending of new generic knowledge and information and the in-depth understanding of the idiosyncratic features of the local production process so as to specify the best applications of the new external knowledge to the characteristics of the production process. At the same time it is the result of the generation of new innovations and the adoption of new capital goods and intermediary production factors, which include a radical reorganization of the vertical division of labor within firms and across industries.

The penetration by the new information and communication technology of the US economy has paralleled a dramatic growth of a service economy which provides knowledge-intensive business services to the rest of the economy by means of advanced telecommunications systems. The vertical disintegration of large manufacturing corporations and the shift towards an immaterial weightless economy are complementary aspects of the generalized introduction of new information and communication technologies. As a consequence the US economy has paved the way towards specific and well-advanced design of the organization of the production process at large where a new division of labor has been put in place with advanced technological and organizational solutions in the whole fabric of the economic system.

Today the adoption of new information and communication technologies in most industries and downstream sectors of activity implies not only the general usage of specific devices to elaborate and transmit information and communication, but also and mainly the specific imitation of the new organization of the production process and the purchase of an array of intermediary products and especially services which have already been tested and tried out successfully in the US economy.

The general characteristics of this new technology are the results of a long-standing process of sequential introduction of systemic innovations, guided by relevant complementarities and interdependencies, mainly elaborated in the US in the last decades of the twentieth century. As such, new information technology reflects the idiosyncratic endowments of the US economy in many respects. This is true both at the aggregate level in terms of abstract input intensities and more specifically at far more disaggregate levels in the fabric of specific applications which have been elaborated in the vertical systems of relations among different fields of application, that is different industries and firms.

The introduction of new information and communication technologies has made possible a substantial increase of total factor productivity levels in the US, far stronger than that of direct competitors because of the appropriate structure of relative prices. The further reduction in relative prices of most productive inputs obtained in the US economy through the 1990s has had further strong effects in terms of competitive advantages based upon the general efficiency of the production process and the even higher differentials of productivity growth of the following sequential wave of incremental innovations characterized by a similar structure of output elasticities. After years of stagnation in the total factor productivity levels the US have in fact experienced a significant increase in the last decade of the twentieth century. This is in contrast to the simultaneous relative decline of the US economy in international markets. US-based firms regained significant competitive advantage and huge flows of foreign direct investments have been attracted by the US economy.

The distinction between shift and bias effects in understanding technological change can provide important insights which explain the foundations of the economics of the vitality of the US economy both in absolute and relative terms.

New information and communication technology today exhibits a strong bias not only in favor of university-based human capital and specific forms of fixed capital, consisting of telecommunications infrastructure, but also an array of intermediary products and especially services which have been sequentially introduced in the US with successful results. At the same time, the general shift effects made possible by the adoption and adaptation of new information and communication technologies are such that almost no country or specific region can resist the new technological opportunity. Adoption, as a consequence, is widespread and diffusion rates are relatively fast with respect to other previous important technologies.

The shift effects provide large opportunities for catching up: laggards and backward countries can take advantage of the new highly productive technologies and reduce the gaps with most advanced countries. The bias effects, however, built into the new information and communication system are such that important differences in actual total factor productivity levels persist over time, because of the differences in the structure of relative prices in imitating countries. The significant requirements in terms of infrastructure and skills act as major barriers to imitation and to entry for most developing countries. Relevant features of the US economy in this respect are a relatively large endowment of academically skilled human capital, a large infrastructure of advanced telecommunication infrastructure and a large and competitive supply of advanced telecommunications services and software.

High levels of variance characterize most countries in the rest of the world both between and within, that is with respect to the US and among themselves, in terms of key relative prices such as the costs and the quality for telecommunications services, software products and most importantly the relative wages of skilled manpower with specific competence in information and communication technology.

The new gaps in international productivity levels and the increased competitive advantage, as experienced by the US economy in the 1990s, can be explained substantially in terms not only of a sheer advantage based upon lead time in the introduction of a new technology, but also and mainly upon the widening gap between actual and potential productivity growth, determined by the specific bias of the new information and communication technology and the sharp difference, with respect to the US, of most competitors in terms of relative prices of key intermediary inputs, human capital and capital goods.

This hypothesis suggests that the shift effect can be offset by the imitative adoption of followers in the short term and as such can provide only a transient competitive advantage to the US economy. The differences in the structure of relative prices, associated with the bias for adopters of the new technology, instead do provide a major scope for a long-lasting competitive advantage for the innovating economy. Only a major effort of followers not only to adopt but to adapt new information and communication technologies to their structure of skills and endowments can reduce such a widening gap. A parallel effort to change the structure of relative prices and to favor the prices of the most productive inputs may help to reduce the production costs and the spread between actual and potential total factor productivity growth.

Because of the complexity of the new information and communication technology system and the tight fabric of requirements in terms of complementarity, interoperability and interdependence among the different technological components, changes in the structure of the technological system are difficult and long-lasting. As such the introduction of incremental and biased technological changes in following countries, aimed at increasing the local fitness to the domestic endowments of production factors, seems most difficult. In turn the change of the structure of relative prices seems a complex process which can last a long time and requires a sophisticated and articulated long-term policy implemented by high-quality policy makers.

Like previous radical technologies, such as the railway, the dynamo, the fordist mass-production system and the plastics and petrochemical industry at large, new information and communication technologies are a bundle of both product and process innovations which apply to a wide variety of activities and relfect the characteristics of the innovating countries and emphasize their structural characteristics. Their international diffusion is pushed worldwide by their strong effects in terms of profitability of adoption and rates of increase of total factor productivity. The consquences of their adoption, however, are far from homogenous across countries. The actual rates of increase of total factor productivity differ systematically across regions and countries, according to their idiosyncratic characteristics and the specificity of their endowments, with respect to the characteristics of the innovating countries (Freeman and Louca 2001).

A large part of the economic history of the last decades of the twentieth century has been shaped by the attempts of followers not only to adopt the new information and communication technologies and possibly participate in the elaboration of the technological system, but also and in some cases mainly as the deliberate attempt to reduce the differences of the structure of relative prices.

The array of privatization and liberation in the telecommunications industry can be considered as the most impressive process set in motion by the advent of new information and communication technologies in the US and by the attempt of followers and prospective adopters to reduce the differences in relative prices of new key inputs.

The specific design of the digital economy as it emerged gradually in the US in the last decades of the twentieth century privileges the role of advanced tele-communications services as the backbone of the new economy. The provision of an array of dedicated, advanced and low-cost communication services is a fundamental condition for the digital economy to spread into the system and for adopters to implement the sequential array of specific tailored incremental innovations which make the system more and more effective.

Divestiture in 1983 and the subsequent Telecommunications Act in 1996 are the basic institutional changes which have paved the way towards such a process in the US. The low prices and high quality of telecommunications services are key conditions for the technological system to keep emerging and for the economy to keep increasing total factor productivity and output growth. The Microsoft case can be considered in a similar vein an important step in reducing barriers to entry and hence market prices for essential key inputs such as software and in general data-processing service and products.

International diffusion of new information and communication technology, made fast by the sheer shift effects, pushes the rest of the global economy to try and reduce the differences in the structure of relative prices with respect to the leading inno-vator. Privatization and liberalization of the telecommunication services industry has become, during the 1990s almost a prerequisite for competitors to enter the new race. The new role of Universities and the provision of skilled human capital is the second achievement brought about by the effort of adopters to reduce the gaps between absolute and actual total factor productivity levels.

Systematic efforts by imitating countries to increase the fungibility of information and communication technologies can play an important role in this context. The localized introduction of contingent technologies able to implement the new higher levels of total factor productivity with a new distribution of output elasticities, more consistent and coherent with the local factors endowments, can reduce the asym-metric effects of the introduction of the new general purpose technology. Such a contingent technological change can take place if firms, in now peripheral countries, are able to apply new information and communication technologies to existing production processes and to existing products, so as to reduce the need to change the techniques, defined in terms of factors intensity, and yet change the technology. The focus on technological blending, based upon the high levels of fungibility of new information and communication technology, can be an important contribution in elaborating both a technology strategy at the firm level and a technology policy at a country level.

12 Policy implications

A European perspective

A narrow definition of economics of innovation yields important insights for the elaboration of a growth-oriented economic policy. The support to both the production and the distribution of technological and scientific knowledge in an economic system is likely to push the rates of growth of output and total factor productivity. A public intervention which favors the accumulation of knowledge and the flows of technological communication within the business community and between the business community and the scientific community is clearly necessary because of the pervasive role of supply and demand externalities in the generation of knowledge, and related problems in terms of limited appropriability, poor excludability and imperfect tradability. Such problems are likely to induce sub-optimal allocation of resources to research and to impede appropriate levels of division of labor and hence efficiency in the production of knowledge. The identification of the clusters of complementary knowledge, both in regional and technical space, and the support to the local dynamics of complementarity, externality and spillovers and hence bounded increasing returns can become an important tool for economic policy in this area.

A narrow definition of economics of innovation focuses the increase of the rates of innovation as the main target for innovation policy. A broader understanding of the interactions between the changes in technology and the changes in the economic structure of the system and the appreciation of the system of inducement mechanisms at work, however, can open up much a larger scope for economic policy. The notions of potential productivity growth, defined as the scope for total factor productivity growth which can be obtained by an appropriate direction of technological changes for given levels of relative prices, and general efficiency, as determined by the reductions in production costs engendered by the appropriate reductions of the relative prices of the most productive inputs, for a given technology, can be used to develop tools for growth-oriented innovation and industrial policy in a global context.

The scope for a growth-oriented industrial policy is evident here. The full understanding of the composition effects makes it possible to develop basic guidelines for an industrial policy aimed at increasing the rates of growth of output in the global economy. The notion of composition effects translates directly into a basic principle to direct the interventions of industrial policy and discriminate across industrial sectors and labor markets.

The stronger the effect of the price of the output of each sector on the output and efficiency levels of other downstream sectors, the larger the scope for an industrial policy intervention. The identification of the sectors providing key intermediary inputs – intermediary inputs which exhibit higher levels of marginal productivity in the downstream production functions – is the first step towards the implementation of an important area of intervention for a growth-oriented economic policy.

It is immediately evident that the position of each industry in the input-output matrix is itself an important factor in directing such an industrial policy. The further above an industry is in the matrix, the larger the direct and indirect effects in the rest of the economy of any induced reduction in the market price for such intermediary products.

At any point in time the configuration of the market price of intermediary products delivered by key sectors can be easily suboptimal. This can occur for a variety of reasons, both in equilibrium and out-of-equilibrium conditions.

Monopolistic markets and barriers to entry in key sectors are a clear and general out-of-equilibrium source of high prices for the products of key sectors. The identification of the priorities for competition policy can be easily derived. Upstream sectors providing key intermediary inputs to the rest of the economy should be the first goal of a careful competition policy aimed at erasing all possible limitations to the full display of the competition process. The persistence of barriers to entry and mobility in these sectors is likely to have much worse effects on the rest of the economy than in any other industry.

Even when full equilibrium conditions are achieved, however, industrial policy can generate higher output and general efficiency levels with deliberate interventions which affect the sectoral composition of public policies. All interventions that are able to change the relative prices of production factors and can ultimately reduce the market prices for intermediary products that are most productive in downstream industries can have positive effects in terms of output and efficiency growth. The parallel change in the fiscal burden and in the levels of duties in favor of key sectors, even if perfectly balanced by an increase in the fiscal burden and duties for non key-sectors, can generate increased general efficiency in the global markets. Moreover, if and when the technological path of innovations seems characterized by a sequence of innovation with similar features in terms of bias, all reductions in the relative prices of the most productive inputs make it possible to reduce the gap between actual and potential productivity growth for the stream of future incoming innovations. Positive effects in terms of average costs are also relevant.

This analysis provides a new rationale by which to assess the opportunity to provide actual subsidies to firms in key industries. Public subsidies to key sectors that can reduce the market price for their products, especially if funded with taxes on non-key intermediary and final products, are likely to engender a strong positive sum game with a multiplier effect that gets stronger as the composition effects in downstream industries get stronger.

The creation of skilled manpower plays a special role in this context. A variety of tools can be applied, such as easier access to academic training, incentives to the immigration of qualified manpower, on-the-job training schemes, increased interactions between the academic and the business community. The reduction of all obstacles to proper mobility into professional jobs is also relevant in this context especially in countries where most professional services are still supplied in markets characterized by high barriers to entry of institutional origins. Support to the creation and full mobility of skills becomes a key issue for countries, regions and industries coping with the transition towards the knowledge economy.

Similar results can be obtained with a change in the composition of the targets of innovation policies. Generally, and in the long term, the analysis of composition effects translates directly into the identification of a clear target as to the direction of technological change. For any resilient composition of factor endowment and relative prices, innovation policies should favor the introduction of technologies that promote the intensive use of the abundant factors in that specific region. In the short term, when the direction of technological change is given and the matrix of input-output flows can be taken into account, it is also clear that a special effort of innovation policies in favor of key sectors is likely to have positive effects if it can increase the productivity of the industries which produce the most effective inputs in downstream sectors. In this case innovation policy can be directed towards the introduction of incremental technological changes in well-defined technological paths which can help the fast increase of total factor productivity levels in an array of downstream industries. The same is true when diffusion policy that is directed to favor the adoption of technological innovations in upstream industries can yield positive effects in the full spectrum of downstream industries.

Innovation policies specifically directed towards the rejuvenation of traditional sectors should be implemented in countries that are forced to adopt and adapt new general technologies that cannot be locally neutral. Such targeted innovation policies should aim at favoring the recombination of new technological knowledge, both embodied and disembodied, with local know-how, and can be based upon a dedicated effort to increase and take advantage the fungibility of the new general purpose technology. Such an innovation policy can yield important positive results, possibly larger that those which might be achieved with the obsessive targeting of the so-called high-tech industries.

Regional policy can play an important role. Economic geography provides much empirical evidence on the strong regional concentration of industries. Industries cluster in a few regions and in turn regions specialize in a few industries. The identification of regions specializing in the supply of key intermediary inputs makes it possible to select them as targets of an array of policy interventions in favor of the production of key inputs. The levels of regional specialization in the production of key intermediary inputs can become an effective tool in identifying priorities and levels of all public interventions across countries in terms of basic infra-structures such as communication systems, advanced education and transportation. Regions provide an important opportunity to implement a labor policy directed to promoting the supply of the specific skills and the kinds of competence requested by new technologies.

The regional dimension of innovation policy is an area of growing concern for economic analysis and policy makers. The distribution of innovation across regions is uneven. European regions differ widely in terms of innovation capabilities. Specifically European regions differ in terms of inputs into innovation activities, outputs of innovation activities and efficiency of innovation activities. The distribution of European funds for innovation across regions is also uneven.

Such asymmetries emerge as a source of problematic concern. From a strict economic viewpoint, regional concentration of innovation capabilities seems the result of strong centripetal forces determined by relevant agglomeration externalities. High-quality skills and institutions (firms, universities and research laboratories) concentrate in a few regions and take advantage of their proximity. Moreover increasing returns, internal both to such regions and to research institutions and firms, are at play. Core regions are better able to generate more innovations, thus making for a more efficient use of available resources. In so doing they attract even more dedicated resources.

The regional concentration of European innovation policy funds reflects such dynamics and is itself a factor of cumulative self-reinforcing processes. Core regions in fact can also take advantage of learning processes in accessing European intervention schemes.

From an aggregate viewpoint this is effective if and when attention is concentrated on the amount of actual innovations facilitated and activated by means of European innovation policies. The specialization of a few regions in innovation capabilities is conducive to increasing the overall amount of innovations introduced and eventually the welfare of the Union at large.

The cohesion of the Union, however, may be endangered. This is all the more true when the effects of other complementary centripetal forces are considered. The increasing levels of economic, social and financial integration of the European countries may act as a major force driving towards increasing concentration of economic activity in a few regions. Factor mobility, in particular, is likely to increase the overall variance in terms of productivity and efficiency levels between core regions and peripheral ones. Relative prices change because of factor mobility, and peripheral regions find themselves trapped into a negative process. The introduction of appropriate innovations is the single most effective tool to counter-balance such dynamics. Peripheral regions need to change and update their technologies in order to face the new relative prices and increase their efficiency and productivity levels.

Here is a major source of contradiction for innovation policy. On the one hand efficiency and self-selection processes favor the concentration of innovation capabilities and innovation subsidies in a few core regions. On the other hand, innovation is most necessary for peripheral regions that are as yet deprived of innovation capabilities and unable to attract European subsidies.

Regional innovation policy at the European Union level needs to develop a framework which integrates two conflicting goals: (1) the needs for innovation policy within peripheral regions and (2) the opportunities stemming from increasing returns within technological districts. The pursuit of the first objective would clearly suggest the maximum of inclusion while the second would require high levels of selection and exclusion. From the strict viewpoint of a research policy agenda the selection of a limited number of technological districts seems to be more appropriate. From a broader economic viewpoint, one which includes the localized context in which technological knowledge eventually generates technological innovations, the second inclusive dimension might yield higher results. The very final word is that only a full and honest awareness of these conflicting goals may lead to the development of a consistent policy agenda. The elaboration, at the regional level, of the distinction between innovation-policies and diffusion-policies might once more be appropriate.

This dual level of implementation should become explicit and intentional. The allocation of resources with auction-based and beauty-contest procedures, now customary, tends to favor the few players located in central regions: they can take advantage of localized increasing returns. Here the evidence gathered stresses the positive role of the close interaction among heterogeneous partners in hastening the rates of accumulation of new technological knowledge and of the introduction of new technologies. A long-standing tradition of intervention in European Union innovation policy towards research consortia might be further implemented. The selection of research projects might include, among other parameters, the local variety of teams. Each team in other words should include more than one member in each site in order to favor the local interaction among universities, research centers and firms.

Second, in identifying the recipients of public funds, the European Union might include, among other criteria, an assessment of the regional location of partners. The positive role of local spillovers might be included in assessing the overall effects of each allocation. Indeed, new emerging empirical evidence suggests that the allocation policy of the resources of the European Union in the area of technological innovation has been extremely selective, with strong effects in terms of a sharp increase in the concentration, at the European level, of R&D activities (Geuna 1999; Antonelli and Calderini 2001).[1]

The identification of basic competencies and their implementation across Europe can become a complementary and yet quite distinct area of intervention. Regions qualify for Community funds when the local infrastructure and human capital capacity happens to be below some threshold levels. Let us assume, for the sake of clarity, that informatics has been selected as a key competence. Let us assume that an appropriate indicator for the basic research capacity in informatics has been developed. Such an indicator can measure the distribution of the key capacity across European regions. Regions which do not achieve a minimum level qualify for specific funds, to be invested in local university and other qualified research institutions.

The consequent distribution of research capabilities across less favored regions can countervail the selective effects of the standard allocation procedures. Universities can become the recipient of explicit policy actions finalized to reduce the variance at the European level, in terms of research capabilities.

A dual-ladder European innovation policy seems most necessary to reduce this diverging contrast. Two actions can be suggested. First, an assessment of the positive effects of local spillovers might be included as parameters in the selection process. Second, a dedicated innovation policy can be designed to focus universities as the recipients of programs aimed at increasing the training and advanced education capabilities in high-tech fields in peripheral regions. Universities can be used as appropriate policy tools to strengthen the local innovation capabilities and favor small and medium-size technological upgrading in peripheral regions when their localized spillover of technological knowledge is enhanced and valorized by means of appropriate innovation policy programs.

The co-evolution of relative prices, technologies and institutions becomes a necessary condition to foster the rates of growth. Institutional change is itself a part of the recursive path-dependent process of change (Nelson and Sampat 2000). Institutions play an important role in labor markets in terms of flexibility and mobility within firms, between firms and across borders; in capital markets, again in terms of mobility and related levels of credit-rationing, within borders and across borders; in the knowledge-intensive interactions that make possible the circulation, and the quasi-trade in technological knowledge among firms and between firms and research centers; in intellectual property rights regimes, in terms of exclusivity and access conditions with important effects regarding respectively enhanced cumulability or increased fungibility; in education and training. Institutional innovation provide the final basket of possible interventions.

All such areas provide the scope for a targeted innovation policy directed towards the introduction and adoption of technological innovations that are appropriate to the local structural, economic and institutional context.

13 Conclusions

A large body of empirical analysis on the main characteristics of technological change in the last decades of the twentieth century confirms that new information and communication technology is at the core of the innovation process and that its introduction – the digital revolution – is engendering a global digital divide. It is more and more evident that the introduction of this new technology into the global economy parallels widening and ever increasing asymmetries among countries and even regions within countries. Such evidence stresses the need for an analytical framework, able to appreciate and integrate at the same time the analysis of the determinants and the effects of both the rates of introduction of new technologies, their direction and their effects.

Economics of innovation has emerged as a distinct field of investigation and specialization in the last decades of the twentieth century. It focuses on the dynamics of new technologies and the analysis of the determinants and effects of the introduction of innovations, mainly at the firm level. Economics of innovation assumes the variety of firms and their heterogeneity in terms of products and factors markets. This approach sees technological change as the endogenous result of the interaction in the market place of heterogeneous agents which are able to change their products and their production functions and can influence, to some extent, the utility functions of their consumers. The analysis of the economic properties of the accumulation of technological knowledge and the actual generation and adoption of new technologies receives considerable attention. In this context the institutional characters of the economics systems in which technological changes are generated, implemented and diffused are investigated with care. Economics of innovation has made possible a far better understanding of the determinants of the rate of technological change and of the role of the competitive arena in the market-place and of the interactions among firms in the assessment of the innovation capabilities of economic systems.

Economics of innovation has emerged as an identifiable area of specialization and systematic theoretical and empirical analysis, as the result of a long-standing process driven by the need to develop a coherent understanding of the evidence for the pervasive role of the residual in economic growth both at the micro and the macro level. The discovery of the residual had proved that technological change is the primary factor able to explain intensive growth. Much attention has been given to assessing the effects of the introduction of new technologies. Yet little analysis had been made to provide an economic understanding of its origins and determinants.

Such evolution has been characterized by a paradigmatic shift in heuristic metaphors, drawn from other natural and social sciences: manna, biological life cycles, Darwinian selection, technological trajectories, networks and systems. Each metaphor led to the development of a consistent sub-set of economic tools: technological opportunities, technological asymmetries, technological path-dependence, technological externalities and technological interdependence. This process has made it possible to build an articulated set of tools with which to analyze jointly the determinants and effects of technological change mainly, if not exclusively, at the firm level.

At this stage a need to recontextualize the main achievements of the economics of innovation is emerging. New technologies need to be analyzed in their economic, industrial and institutional context of adoption. A new technology can be more productive and efficient in one context rather than in another. A new technology can be actually superior in one context and less efficient than existing ones in other contexts. The role of the relative prices and of the institutional context that characterizes each economic system needs to be properly appreciated. Some systems are better able to generate some kinds of technological knowledge and changes than others. The issues of technological variety seems more relevant in the new global economy.

By the same token the need to fully appreciate the context of generation and introduction of the new technological knowledge and the new technologies is clearly relevant from an innovation policy perspective. Appropriate technologies exist and they should be identified and become the target of selective innovation policy tools.

The merging of the economics of innovation and the economics of technical change can provide a broader and richer analytical framework which is far larger than its single components. Much less attention has been paid by the economics of innovation to the structural characteristics of the systems in which such processes take place. Too much attention has been paid in the economics of innovation to the analysis of the rates of introduction of new technologies. And too little attention has been given to the direction of technological change and its relations with the structural characteristics of the context of introduction. The analysis of the interactions between the direction of technological change and the relative prices of production factors and their industrial and economic determinants is left out of the central scope of analysis. This constitutes a substantial analytical weakness and limits the capability of the economics of innovation to provide an overall understanding of the dynamic properties of economic systems as a whole.

The economics of technical progress, as developed since the 1930s and through the 1960s, instead, addressed primarily the analysis of the structural effects of the introduction of new technologies. The analysis of the main characteristics of technological change, whether labor- or capital-intensive, and the consequences on output and efficiency of their introduction had been mainly developed within the context of a general systemic analysis very much oriented towards the economics of growth. The economics of technical progress was mainly devoted to the analysis of the introduction of a single specific new technology into a simplified economic system characterized by single factors markets. The economics of technical progress had little in the way of microeconomic foundations and the actual dynamics of the accumulation of new technological knowledge and the generation of new technologies, at the firm level, was left out of the scope of the analysis.

After much progress in the economics of innovation, it seems now necessary to build a bridge towards the main achievements of the economics of technical progress and to try and retrieve the main results achieved in that context, into a different and broader context of analysis, one where heterogeneous firms are at play in a variety of factors and products markets and coexistence among and between old and new technologies at each point in time is taken into account.

A satisfactory analysis of technological change must integrate the understanding of the determinants and effects of both the rates of introduction of innovations and of the direction of the new technologies being introduced. Indeed, at the system level, relative prices and technological change interact with significant effects. Relative prices have a direct bearing on the assessment of technological change both with respect to the spectrum of techniques of each technology and with respect to the profitability of one new technology with respect to another.

The direction of technological change can affect the rate and vice versa, when a complex economic system where a variety of regions with different endowments and different relative prices of production factors is allowed and their change in time is taken into account. Because the ranking of new technologies depends upon the relative prices of production factors, technological reswitching can take place when the relative prices of production factors change.

Two important notions have been introduced. First, the distinction between potential and actual total factor productivity growth. Potential total factor productivity growth is obtained, in a non-neutral production function, when the most productive input is cheapest. Second, the distinction between general efficiency and total factor productivity growth. Production costs, for a given technology, are influenced by the levels of relative inputs costs. The general efficiency of production can increase not only because of the introduction of a new technology, but also by means of a reduction of the relative price for the most productive input. The effects of given relative factors prices on the range between potential and actual total factor productivity levels and the consequences of the changes in the relative inputs prices on production costs, for a given technology, can be termed composition effects.

In this context the appreciation of the role of the direction of technological change and the context of its introduction make it possible to better appreciate the distinction between general technological change and contingent technological change.

General technological change is characterized by innovations which are – at least locally – Hicks-neutral and make possible shifts of total factor productivity levels such that the new isoquant lies everywhere at the left of the old one. Contingent technological change consists just in a bias effect and is only locally progressive (and regressive). These notions can help explain much puzzling empirical evidence which came into consideration in the economics of innovation and new technologies. The notions of general and contingent technological change make it possible to complete the full mapping of technological change, in a broader systemic context where the variety of product and factor markets is now taken into account and consequently the issue of technological variety is considered beyond the deterministic assumptions about delays in diffusion processes.

The identification of the notions of general and contingent technological change can be considered an important contribution to the more general analytical merger between the economics of technical progress, much practised since the 1930s and through the 1960s and the area of investigation, frequently termed economics of innovation, which has emerged since the seventies. The economics of technical progress focused on the analysis of the characteristics of the system in conjunction with the introduction of single new technologies. The emphasis was mainly put upon the relations between the theory of distribution, the markets for production factors and the consequences as to production.

The economics of innovation has a stronger microeconomic foundation and focuses mainly on the behavior of firms and the economics features of the accumulation of knowledge, the introduction and diffusion of new technologies, paying special attention to the dynamics of local externalities, interdependence and complementarity among technologies and firms. In the economics of innovation, technological change is viewed as the endogenous outcome of the out-of-equilibrium interaction of producers and consumers in many products and factors markets that are characterized by relevant heterogeneity and intrinsic variety of agents. In the economics of technical change the attention was focused upon the dynamic implications of the introduction of single innovations in a single economic system.

Many results of the economics of technical progress could not easily be applied to the field of investigation of the economics of innovation for the lack of appreciation upon the effects of the interaction in global products markets of agents based in a variety of factors markets. The evidence of the new global economy makes this context relevant.

The debate over the classification of techniques absorbed much attention in the effort to assess their effects in terms of factor demand as well as the efforts to classify technological change in terms of direction as measured by the changes in factor intensities shaped by the new forms of the production process. Little effort has been made to combine these two debates and use their outcome so as to develop analytical tools useful for the economics of innovation and new technology in a broader and dynamic out-of-equilibrium framework.

The merging of the economics of innovation and the economics of technical progress seems more and more necessary and promising. It is promising because it provides a systemic field of investigation where many advances of the economics of technical progress find new support and a broader context of application and cumulative exploitation. It is necessary because it is an important and promising way to contrast the progressive reduction of much current analysis in the area of economics of innovation to a theory of the innovative firm; this analysis is often carried out with little and even declining attention dedicated to grasp the structural factors which in fact shape and assess the actual outcome of the introduction of an innovation.

Indeed, many innovations are introduced at each point in time and only a few are actually adopted and implemented. The structural characteristics of the system make the difference among innovations. The understanding of the selection context for innovations is necessary to understand the actual rate and direction of technological change. Without such a context, much economics of innovation, after forty years of successful growth, might risk declining into a subjectivist theory of cognitive creativity. In this case it could deliver little more than an ever changing recipe book for animal spirits.

The merging of the economics of innovation with the economics of technical change makes it possible to retain the deep understanding of the Schumpeterian out-of-equilibrium context in which technological change takes place and to generalize it in a broader analytical framework. Such an attempt is more and more necessary when much recent theorizing the new growth theory pretends to provide a dynamic equilibrium approach based upon smooth and steady rates of introduction of new technologies.

The merging of the economics of innovation tradition with the economics of technical change makes it possible to reappreciate the strength of the inducement approach to explaining the determinants of the rates and the direction of technological change. In this context the decoupling of the inducement to innovate from the inducement of the direction of technological change emerges as a fertile area of investigation.

The structure of relative prices becomes a key factor to assessing the actual effects and determinants of the rate and direction of technological change when the variety in factors markets is take into account. Indeed, the new global economy, as it emerged in the last decades of the twentieth century, is more and more characterized by single and integrated international products markets where firms located in a variety of different factors markets are engaged in the same competitive arena. This coexistence of single product markets and a variety of factors markets can be considered to be the distinctive feature of the international economy at the end of the twentieth century. In such a context the analysis of the interplay between the rate and the direction of technological change and the relative factors prices of each local market can no longer be regarded as a scholastic exercise and becomes relevant in understanding the actual dynamics of both firms and economic systems.

An important distinction has emerged between internal and external path-dependence. Internal path-dependence takes place when the path along which the firm acts is determined by the irreversibility of her production factors and by localized learning. External path-dependence is instead determined by the external conditions. In the first case the introduction of the new technology is influenced by the switching costs firms face when they try and change the levels of their inputs. In the latter case, on the other hand, the introduction of the new technology is shaped by the factors and products market conditions.

The notion of path-dependence developed by Paul David (1975) and subsequently implemented (Antonelli 1995) belongs to the first case: firms are induced to follow a path of technological change by their internal characteristics, including the localized learning conditions. The notion of path-dependence elaborated by Brian Arthur (1989) and Paul David (1985) clearly belongs to the second case: new technologies are selected by increasing returns to adoption at the system level. The appreciation of the role of technological externalities in the generation of new technological knowledge and in the introduction of new technologies has contributed to the analysis of external path-dependence (Antonelli 1999a and 2001a). The model elaborated in this book provides a synthesis of the notions of internal and external path-dependence. Internal path-dependence is determined by irreversibility, localized knowledge and switching costs that are specific and internal to each firm. External path-dependence is shaped by the role of factors endowments and relative prices that induce the direction of technological change.

The distinction between path-dependent innovation and path-dependent diffusion becomes relevant in this context. The analysis carried out in this book focuses on the path-dependent aspects of the localized introduction of new technologies. In so doing it differs and yet complements the analysis of the path-dependent characteristics of the diffusion of rival innovations elaborated by Arthur (1989) and David (1985). In the former analysis the selection and eventual diffusion of new technologies is path-dependent in that it is influenced by the timing of their sequential introduction which in turn affects the relative profitability of their adoption because of the powerful consequences of positive feedbacks consisting in network externalities and increasing returns in production. In the analysis here, instead, path-dependence consists in the mix of irreversibility, induced innovation, local externalities and local endowments. The contrast between the irreversibility of the tangible and intangible stock of sunk inputs and the actual conditions of both factors and products markets, affected by the continual introduction of unexpected innovations in the system, is the prime engine. This contrast induces the creative reaction and the eventual introduction of new localized technologies. Myopic, but creative, agents introduce technological changes that are localized by their knowledge-base built upon localized learning processes, the switching costs stemming from irreversible production factors and the external conditions of factors markets. While the rate of introduction of new technologies is induced by the contrast between the irreversibility of production factors and the actual conditions of the markets, the direction of the new technologies is induced by the relative prices of production factors. The access conditions to technological spillovers and external knowledge at large affect the actual results of the induced innovation activity of each agent.[1]

In this approach the variety of actors and interacting markets matters. Firms are induced to innovate when their myopic expectations do not match the actual markets. Firm, in other words, react to all changes to their myopic expectations, not only with changes in the price-output mix but also with the introduction of new technologies. The new technologies reflect the specific context of action. Such a context includes firms active in a broad industrial structure which includes regions and countries with significant heterogeneity in both technologies and endowments. The new global economy provides a context in which an international industrial dynamics approach seems necessary.

The introduction of general purpose technologies with a wide spectrum of applications to a variety of products and production processes jeopardizes, at each point in time, the traditional system of comparative advantages upon which symmetric trade among agents located in heterogeneous factors markets can take place. In international economics, heterogeneity of factors costs can be compensated for only by technological variety. Countries and regions where the supply of labor is abundant and the supply of capital relatively scarce should use labor intensive technologies and, vice versa, capital-abundant countries should use capital-intensive technologies. The reduction of technological variety brought about by general purpose technologies may cause serious problems for the complementarity of regions in the market-place and the symmetric distribution of gains from trade in the global economy. Economic history provides, indeed, a large body of evidence on the negative effects for peripheral regions of the introduction in core regions of general purpose technologies which, at least in a first and yet long phase, reduce technological variety to a single dominant technological design where the characterization of the technology reflects more the endowments of core regions than those of peripheral ones (Freeman and Louca 2001).

The ranking of regions can be actually be derived in terms of the mismatch between endowments and output elasticity of production factors. The relationship between the direction of technological change and the local structure of endowments leads to a definition of economic centrality. Core regions are those where the mismatch is nil, while in most peripheral ones, locally abundant factors exhibit the lowest relative levels of output elasticity.

Thus it is clear that the direction of technological change plays a crucial role and an effort should be made to better understand the determinants of the factor intensity of new technologies.

Technological change and structural change are complementary and interdependent aspects of a broader and systemic process of path-dependent evolution. The structure of the system shapes the direction of technological change which in turn affects the structure of the system. The changes in the structure of the system induce the introduction of further innovations. The process is endless, as is the transition from one potential – and yet never achieved – equilibrium to the next, along a path shaped by out-of-equilibrium feedbacks and recursive causation where the path provides the record as well as the direction.

This analysis provides a structural context in which the achievements of economics of innovation with respect to the production and distribution of technological knowledge, the determinants of innovative choice and the effects of the introduction of innovations on industrial dynamics can find an appropriate use. Not only the rate but also the direction of technological change is endogenous to the interaction of market forces.[2]

The new global economy provides a new specific context in which the innovation activities of firms are embedded. Such a context provides important incentives and constraints to the innovative activity of firms which deserve to be carefully assessed and understood. In so doing, economics of innovation acquires a broader analytical prospect and can integrate the analysis of the interactions between the structure of economic systems and the aggregate effects of the rate and direction of technological change.

The basic assumptions of economics of innovation as to the endogeneity of technological change remain at the core of the analysis. The understanding of the constraints and incentives to introduce and adopt new technologies, however, can now take into account the broader conditions of the economic system as a whole and in this context can better appreciate the specific conditions which shape the innovation strategies of the firms and their possible effects at the aggregate level.

The introduction of general and contingent technological changes may be considered as the outcome of specific and intentional strategies and competencies. At the microeconomic level it now seems clear that any changes both in the shape and in the position of the available isoquants have strong and asymmetric effects on the performance of heterogeneous firms. At the system level we see that rigid systems of relative prices can direct endogenous technological change.

The structural architecture of the economic system, namely the composition of industrial sectors, the characteristics of their markets, and their location in diverse factors markets, and the vertical relations among them play a key role in assessing overall performances. Contingent technologies cannot be imitated by firms active in regions and industries where the structure of relative prices differ substantially from those of innovators. Contingent technological changes do not diffuse in different contexts and yet they help in increasing total factor productivity levels and reducing costs with evident effects in terms of barriers to entry. Contingent technologies lead to absolute costs asymmetries among firms.

The dynamics of flexible relative prices can explain the diffusion paths of the new technologies. Regional integration of economic systems and consequent enhanced factor mobility between regions characterized by different, absolute and relative, factors costs and by the use of different technologies may yield negative consequences in terms of a reduction in the general efficiency of the production process, as measured by the levels of production costs. These effects can be counterbalanced only by means of the enhanced rate of introduction of new technologies. The introduction of a new technology may have important effects in terms of discontinuity in factors markets with major macroeconomic imbalances. International competitiveness is strongly affected by technological change and with different effects whether the new technology is general or contingent. Industrial policy can play a major role in shaping relative prices especially when intermediary inputs are taken into account.

The rates of growth of productivity and efficiency and hence the opportunities for growth in the global economy at the system level depend on the key sectors, the industries that feed the rest of the system and provide new and highly productive inputs to the other industries.

The need and the relevance of this international industrial dynamics approach where both the rate and the direction of technological change are induced by disequilibrium conditions, seems especially clear and strong in the new global economic system and the new wide and heterogeneous regional integrated system such as the enlarged European Union. The new global economy exposes each economic system to a new kind of competition where firms located in different factors markets, with different factors prices, compete on quite homogeneous global product markets. In this context countries, and within countries regions, differ widely in terms of their ability both to generate new technologies and to take advantage of them. The evidence on the great and ever increasing gaps among economic systems and regions within each economic system despite the increasing mobility across borders of technologies and products requires an interpretative framework which is able to appreciate the role of the co-evolution of technologies and economic structures.

The analysis of the interaction between relative prices and the rate and the direction of technological change yields three important results. First, it makes it possible to sharpen the distinction between contingent and general technological change. Second it sheds a new light on the induced technological change approach. Third, it provides some ways of understanding the recursive and path-dependent causation between technological change and structural change. Let us consider briefly these issues in turn.

The distinction between the inducement of innovations activated by the changes in the levels of the relative prices of production factors and in the levels of the demand for products, and the inducement of the direction of technological change determined by the levels of the relative prices for production factors, seems important because it provides a coherent and articulated system of complementary inducement mechanisms. Changes in the relative factors prices and demand for products lead to disequilibrium conditions. Firms can react either by changing their techniques or their technologies. Irreversibility and switching costs induce firms to change their technologies. The levels of relative prices, as opposed to their changes, induce the direction of the new technologies, i.e. the more intensive use in the production process of the production factors which are locally more abundant and less expensive. Composition effects induce the direction of technological change.

Production costs and hence output levels in global markets are sensitive to changes in factor intensity, as determined by both the changes in the relative costs of production factors and in their relative marginal efficiency. Such composition effects become apparent when the relative scale of the production factors change. Composition effects have major consequences in terms of technological variety. Technologies cannot be longer ranked according to total factor productivity levels only: the marginal productivity of inputs and their relative costs in factor markets matter in assessing actual productivity levels. Composition effects affect standard procedures of assessing total factor productivity growth.

Substantial differences between actual and potential efficiency arise and the ranking of technologies is contingent, as it is strongly affected by the factor bias and the relative costs of production factors. Moreover for any given technology, production costs are affected by relative factors prices and such effects are stronger, the stronger the difference in the output elasticity of production factors. Such differences are most relevant at a desegregate level of analysis and in an analytical framework where a variety of firms and industries, active in a variety of local factor markets, and a variety of specific processes of introduction of new technologies, is taken into account. The traditional methodology of analysis of total factor productivity, based upon the total differential, is biased by the lack of appreciation of the composition effects. This is not the case for an analysis based upon the cost function. In the cost function the wage to capital rental costs ratio becomes explicit and the effects of changes in the unit costs of production factors on average total costs can be easily accommodated.

Changes in relative prices of production inputs, especially when intermediary factors are taken into account, are often determined by increasing globalization. In these circumstances the ranking of technologies is fuzzy and technological variety emerges both synchronically and diachronically, with evident problems both for economic analysis and managerial decision-making. Substantial technological variety emerges from our analysis as a result of the composition effect, the sensitivity of output and average costs to the system of relative factor prices contingent upon their relative marginal productivity. The choice and adoption of new technologies is contingent upon the relative prices and the output elasticity of production factors.

The recursive causation between technological change and structural change now becomes apparent. The introduction of new technologies is likely to affect the structure of the system, in terms of relative and absolute costs of production factors and skilled labor. Such changes in turn lead to higher levels of disequilibrium conditions. Disequilibrium cum irreversibility and switching costs induce the introduction of new technologies. The direction of the new technologies selected in the market-place is induced by the levels of relative factors prices. The introduction of new, locally neutral as well as contingent technologies, has direct effects on competitors, localized in other factors markets and for downstream and upstream customers and suppliers. This generates new disequilibrium which feeds back inducing further technological changes. The process is endless and explains much of the dynamics of growth and change in disequilibrium conditions the historic evidence provides.

These issues seem most relevant in the current analysis of the rate, direction and effects of technological change. The rapid technological change associated with the introduction of new information and communication technologies is clearly biased. New information and communication technology has all the characteristics of a general purpose technology which makes possible a sharp increase in total factor productivity for a wide range of relative factors prices and yet has a strong bias in terms of output elasticity which reflects the specific path of implementation and introduction. Such a path is embedded in the characteristics of the innovating countries and it is coherent and consistent with the local endowments and the local processes. Their widespread adoption and diffusion in the global economy is necessary and even compelling because of the quantum jump in productivity they make possible and yet it can be the source of major and ever increasing asymmetries among different countries with different factors endowments where different technologies with different bias were at work.

New communication technologies are severely biased. They are highly capital- and skilled-labor-intensive and by contrast unskilled-labor-saving. Second, and more important, the adoption of information and communication technologies parallels the strong increase in the use of advanced telecommunications services. Finally, new communication technologies can be characterized as a sequential stream of new complementary technologies aligned along similar characteristics in terms of the prevailing output elasticity of production factors. Advanced telecommunications become a key sector: the supply conditions and the relative prices of advanced telecommunications services are key to assessing the actual productivity and efficiency in the rest of the system.

Industries located in regions where skilled labor is abundant and the supply of advanced telecommunications services is efficient and innovative are likely to benefit most from such a direction of technological change. Their effects in industries and firms located in regions where the supply of advanced telecommunications services is inefficient, unskilled labor is abundant and both fixed and human capital is scarce, however, are likely to be much less strong.

The effects of globalization in this context are likely to be most important in terms of a widening gap in terms of actual efficiency and hence competitivity, market shares and rates of growth across countries. The reduction of technological variety activated by the diffusion of new information and communication technologies can put at risk the source of comparative advantages for heterogeneous countries, where heterogeneous technologies were used. A hierarchy of industries across regions is likely to emerge, one where industries located in skilled-labor-abundant countries and where advance telecommunication services are supplied in an effective and competitive market place, are able to reap most advantages.

The new direction of technological change brought about by the wave of innovations in information and communication technologies is likely to have radical composition effects with significant effects in terms of technological variety and the resilience of older technologies according to their bias and the local factors markets. Composition effects are likely to lead to the identification of a strong hierarchy of industries and regions in terms of actual efficiency, depending on the relative costs of skilled labor and advanced telecommunications services. The interaction of the composition effect, and the levels of total factor productivity and general efficiency, is relevant not only synchronically – across countries and regions – but also, and worst from an analytical viewpoint, diachronically. One technology can rank better than another technology, in a given country and for a given system of relative prices. When the latter change, however, the technological ranking may be reversed and the inferior technology actually becomes better and vice versa.

Highly specific technological paths are likely to arise in each local factor market and in the long term because of the interaction between technological change and composition effects. The direction of technological change can be considered endogenous both because of the selection that each local system will make in terms of adoption of technologies better suited for the local system of relative prices and because of the incentives to generate and introduce innovations that are not only potentially but also actually more performing in each specific factor market condition.

Next to heterogeneity of factors markets, heterogeneity of consumers can play an important role in this context. The tastes and preferences of consumers differ across regions. The introduction of product innovations in well-defined product niches, dedicate to groups of consumers, can secure a fraction of the general demand curve for well-identified products. In this case product innovation can generate relevant mark-ups which balance the cost asymmetries emerging from the mismatch between the bias in technologies and the local endowments. Heterogenity of profit rates across niches and regions, together with technological variety, is the ultimate effect of the introduction of the hypothesis of basic heterogeneity in consumer tastes and factors endowments.

The introduction of niche-product innovations and/or of process innovations directed to generate properly biased technologies, able to minimize the mismatch between local endowments and output elasticities of production factors, become the two prongs of a locally aware technology strategy.

The relationship between economic integration and the rates and direction of technological change can now be assessed. According to a large literature characterized by substantial technological determinism, globalization is the result of new technologies. New information and communication technologies are considered a main factor of globalization because of their strong effects on communication costs, the generalized footloose character and the strong centrifugal consequences of their adoption in terms of dispersion of economic activity across regions.

According to our analysis, however, the direction of the analytical arrow can be reversed. Rapid technological change can now be considered as the consequence of economic integration, often decided at a political and institutional level. Economic integration has put the competitive equilibrium of disparate countries at risk for the increased mobility of production factors and goods across regions. Regions needs to increase their productivity in order to survive in the new global market-place. Such an increase in actual productivity can take place in three ways: (1) Countries can change the factor intensities of their technologies according to new conditions of local factor markets. (2) Countries can increase the general efficiency of the production functions. (3) And finally countries can adjust their relative prices to those of innovating countries and adopt their technologies. Now the rapid rates of technological change experienced at the end of the twentieth century can be considered to be a result of the increased levels of globalization of the world economy.

The hypothesis that the last decade of the twentieth century, mainly in the US, has been shaped by the introduction of a general technological change, characterized by the increase of total factor productivity associated with significant capital and skill-augmenting effects, seems consistent with the characteristics of the information and communication technologies whose implementation requires major investments in fixed and human capital. In these conditions the economic analysis of the effects of the transition to new technology, at the system level, takes on a more realistic flavor which helps in elucidating the present combination of continuous introduction of new technologies associated with discontinuous structural change, both with respect to the relations among industries within the US economy and especially with respect to the sharp increase in gaps and asymmetries among trading partners and between the US economy and their main trading partners.

Technological change, in this perspective, engenders significant asymmetries in international markets. The introduction of innovations confers transient competitive advantages based upon lead times of innovators and imitation lags of adopters, when the latter have access to the same factors markets. Technological change also engenders long-lasting competitive advantages based upon differentiated levels of total factor productivity levels between innovators and adopters when factors markets differ. Finally, for any given technology, the competitive advantage of nations is influenced by the changes in the relative prices and their consequences in terms of production costs. In turn both transient and long-lasting asymmetries induce new waves of technological change in an endless process, which all takes place far away from equilibrium. The current efforts of the "new growth theory" to portray long-term growth as a smooth equilibrium process where no structural change takes place, with no interaction between the introduction of new technologies and the structural characteristics of the economic systems, are at odds with the results of this line of analysis (Romer 1986, 1994; Aghion and Howitt 1998).

The continual interaction between endogenous technological change and the structural dynamics of the system takes place in out-of-equilibrium conditions where agents able to manage the coevolution of both technologies and relative prices have better performances. The sharp effects in terms of discontinuities in factor markets such as the fall of the demand for labor and the sharp increase in the demand for capital and the parallel strong surge in real interest rates, stimulated as they are by small changes in the ratio of wage to capital rental costs, call attention to the criticality of parameters in the systems.

Second and more important, it seems clear that such discontinuities, endogenously generated by the continuous dynamics of technological change, can be too drastic for the system to be able to accommodate them via the traditional compensation effects. Out-of-equilibrium conditions are likely to emerge and last a long time before agents are able to reorganize their expectations and learn how to recombine the new labor productivity levels with prices and investments rates. Product prices stick to the previous higher levels for the classic inertia of oligopolistic markets and reduce the speed of reaction of the system towards the new equilibrium. Indeed, it is well known that all adjustments of product prices and capital stocks require a long time to be elaborated. Only when product prices actually drop, according to the increased levels of total factor productivity and the levels of inputs prices, and new investments are made according to the new levels of capital productivity and the prices of capital goods, can the sharp reduction in employment be absorbed back into the system. Long-term unemployment can be a consequence of a continuous effort of introduction of capital augmenting incremental technological changes in a macroeconomic context shaped by the lack of any measure to sustain overall demand levels. Schumpeterian growth cycles associated with the introduction of new radical innovations can be explained in terms of the market dynamics in upstream markets for production factors. The economic effect of new biased technologies in terms of productivity and output growth is shaped by the dynamic interplay between the new derived demand for production factors and their supply in intermediary markets. Simple assumptions about entry and exit in such markets explain the time path of productivity and cost-efficiency and hence opportunities for output growth in downstream industries.

The hypotheses about a growth dynamics shaped by punctuated phases of introduction of new technologies and punctuated swings of economic growth, developed in the neo-Schumpeterian tradition, seem complementary to our analysis. This is true especially when an historic trend in the increase in real wages is assumed as an underlying dynamic factor and firms located in advanced countries where human and fixed capital are abundant, are expected to be able, by means of learning processes, to influence the direction of technological change according to their own factor endowments (Abramovitz 1989; Mokyr 1990; and Antonelli 1999a).

This analysis is important on many counts. First it makes clear that for each endowment of production factors there is a best technology and vice versa. The coevolution of the direction of technological change, towards the introduction of a bias which pushes towards the intensive use of abundant and hence less expensive factors and vice versa the evolution of the endowments, or at least relative prices, that make the most productive inputs least expensive, can yield important positive effects in terms of total factor productivity and output growth. The divide between actual and potential levels of total factor productivity growth, production costs and hence output depends in fact on the relationship between the specific system of relative prices in each factor market and the direction of technological change. Second and more important, it provides a strong foundation to the understanding of the role of key sectors in industrial dynamics. Key sectors are those industries that supply the most productive intermediary inputs to the rest of the economy. Third, Schumpeterian business cycles can be the rational result of learning processes in entry and exit and related structural change respectively in the sectors producing the most and least productive intermediary inputs. The economic effects of regional integration can be better assessed and the effects of the changes in the relative prices can be appreciated.

Finally, a new agenda for economic policy can be built around the notions of general efficiency in terms of production costs and of the range between actual and potential productivity growth, defined as the scope for measured productivity growth which can be obtained by appropriate levels in the system of relative prices. Relative prices of key inputs become the clear target for economic policy. Industrial policy should focus the industrial dynamics of such key sectors to make entry and supply as elastic as possible so as to reduce monopolistic rents and delays in the final adjustment to competitive prices. Innovation and regional policy can target the industries that produce the most productive intermediary inputs. The distribution of fiscal levies across intermediary products should take into account the downstream composition effects of the relative prices. Education and training policies play a key role here as tools to accelerate the provision in the labor markets of the skills which are complementary to the new emerging technologies.

The results of our analysis are most important from a regional innovation policy perspective. The central problem for regional innovation policy seems to be whether the emerging centripetal forces in the distribution of innovation capabilities within the European Union should be accommodated in order to increase the overall rate of technological advance or counterbalanced with dedicated innovation policy programs directed towards peripheral regions.

These elements are all the more relevant when we recall that the results of many empirical analyses of the present rate and direction of technological changes which are being introduced confirm their human and fixed capital-augmenting characteristics as well their strong effects on total factor productivity. This evidence and the qualitative analyses of the pace and implementation of new information and communication technologies suggest the need to reconsider that debate and try and extract some workable implications, in the well-defined context of the economics of innovation and new technology.

The analysis of the stylized facts of the new rate and direction of a technological change based upon much case-study evidence suggests that a new phase of technological change has been taking place in the last twenty years. New technologies are more and more centered upon the emerging technological system based upon new information and communication technologies. The new technological system is emerging gradually as the result of a myriad of innovations being introduced and eventually made complementary by market forces, by means of incremental selection, implementation and adoption. The new technological system based upon new information and communication technologies require very high levels of fixed capital infrastructure and make use of highly sophisticated labor. The introduction and diffusion of new information and communication technologies parallels the radical change in the organization of production processes. Services now account for more than 60 percent of gross national product in most advanced countries and the knowledge contents of products is ever increasing. The new economy is also and substantially a knowledge economy. A variety of transitions paths to the new knowledge economy can be detected. The adoption and recombination of new information and communication technologies leads to a myriad of specific new products and processes. This process takes place generally in a context where a great variety of factor markets must be taken into account (Brynjolfsson and Kahin 2001; David 2001; Antonelli 2001a).

Empirical evidence elaborated at a more aggregate level suggests that the present rate and direction of this technological change is characterized by continual rates of increase of potential total factor productivity and a significant factor-augmenting effect in terms of an important increase in the output elasticity of both fixed and human capital stocks and a significant decline in the output elasticity of poorly qualified unskilled labor (Abramovitz 1989). The analysis of macroeconomic evidence shows that the introduction of new technological systems, based upon a continuous introduction of a sequence of incremental innovations in information and communication technologies, aligned along a well-defined set of similar characteristics in terms of factor bias, has been often accompanied, especially in followers, by massive and structural unemployment.

This set of conditions provides a context in which the identification of the distinction between general and contingent technological change, the analysis of technological variety, i.e. the coexistence of locally progressive and regressive technologies and the assessment of technological substitution, that is the transition from one set of techniques to another induced by changes in factor prices, can become a valuable tool of analysis for both economics and policy-making.

Finally, the analysis of the full set of dynamic and causal interdependence between structural change and technological change makes it possible to grasp the path-dependent and recursive character of economic growth. Innovation is no longer regarded as an exogenous or autonomous process, but as the endogenous result of a well-defined set of incentives and constraints built into a recursive inducement mechanism where the changes in the structure and in the technology feed each other. New technologies are introduced to face the disequilibrium generated by structural change with a direction defined and selected by the relative abundance of basic and intermediary inputs. The introduction of new technologies changes the economic and industrial structure of the system and hence generates new disequilibrium conditions. This in turn feeds back, stimulating the introduction of further innovations. The path-dependent laws of motion of the system are now set.

Notes

2 Shifting heuristics in the economics of innovation

1   The inducement hypotheses deserves much more attention than it has received after the first interest. See Arrow (2000) and Chapter 4 for a reappraisal.
2   See Chapter 4 for a more explicit analysis.
3   A quote from the recent remarkable book by Richard Caves helps to grasp the shift in the understanding of such dynamics: "Economists usually assume that the competitive firm's primary concern is to protect proprietary knowledge from appropriation by would-be raiders. Yet the firm's better strategy may be to tolerate extensive leakage of knowledge from its corridors and conference rooms, in exchange for keeping its own receptors tuned to knowledge seeping from competing firms. Extensive swapping of information between employees of competing firms takes place in many high-tech activities, as indeed job-hopping from firm to firm" (Caves 2000: 367).
4   According to Scherer (1999) the analytical core of the new growth theory consists in three steps: first, the identification of two quite distinct components of knowledge, the generic and the specific; second, the new understanding of the interaction between the two components of knowledge; third, in the assumption that such interaction leads to increasing returns with the classic spillover of knowledge externalities: "The human capital is made more productive by interacting with the stock of knowledge which includes knowledge of all designs previously achieved along with the scientific knowledge published by academic researchers. The more knowledge there is, the more productive R&D efforts, using human capital are" (Scherer 1999: 35).

3 The retrieval of the economics of technological change

1   See Robinson (1937: 139): "In a discussion of the effects of changes in techniques upon the long-period equilibrium, in my *Essays in the Theory of Employment*, I made use of Mr. Hicks' classification of inventions, according to which an invention is said to be neutral when it raises the marginal productivities of labour and capital in the same proportion, and it is said to be labour-saving or capital-saving according as it raises the marginal productivity of capital more or less than that of labour, the amounts of the factors being unchanged. I analysed the effect of an invention upon the relative shares of the factors in the total product, when the amount of capital is adjusted to the new technique (so that full equilibrium is attained, with zero investment), in terms of this classification of inventions and the elasticity of substitution, showing that, with a constant rate of interest, the relative shares are unchanged, in equilibrium, by an invention which is neutral in Mr. Hicks' sense provided that the elasticity of substitution is equal to unity, while if an invention is labour-saving or capital-saving in Mr. Hicks' sense, the relative shares are unchanged (in equilibrium, with a constant

rate of interest) if the elasticity of substitution is correspondingly less or greater than unity. Mr. Harrod made some criticisms of my analysis which lead to the suggestion that it would be more convenient to use a classification in which an invention is said to be neutral when it leaves the relative shares of the factors unchanged, with a constant rate of interest, after the stock of capital has been adjusted to the new situation."

2   Elaborating along these lines significant progress has been made by scholars specializing in the analysis of the international transfer of technological change, such as Westphal (1990), Evenson and Westphal (1995), Lall (1987), Teitel (1987) and Katz (1987). According to this line of enquiry the context of adoption of a technology plays a key role in assessing its actual efficiency. The application of a given technology to a specific context moreover requires much effort so that the line between adoption and incremental innovation is blurred. The argument elaborated in this book complements this approach, yet stressing the key heuristic role of the differences in relative prices.

4 Composition effects

1   The analysis will consider a simple two basic factors production function for the sake of clarity.

2   The distinction between contingent technological change and biased technological change becomes clear here. A technological change is contingent when it is biased and moreover it is characterized by a (small) shift such that it engenders a total factor productivity growth only within a limited range of possible relative prices of production factors. Conversely, a technological change is not contingent when either it is neutral, or it combines both a strong shift effect and a small bias. In any case a technological change is general when its application, in all possible regional factors markets, is likely to engender an actual increase in total factor productivity levels.

3   See Ruttan again: "An implication is that the gains from labor-saving technical change are less in a low wage than in a high-wage economy. What happens to index number bias when non-neutral technical change is combined with changing relative prices? Suppose that the factor-saving and price effects both act in the same direction as when "labor-saving" technical change is combined with increases in the price of labor relative to capital? In this case the rise in the price of labor induces substitution of capital for labor and the technical change induces labor saving by increasing the marginal productivity of capital relative to labor. In this case the index number bias and the neutrality effect tend to be cumulative . . . Suppose, however, that the factor saving effect and the price effect act in the opposite direction (technical change is autonomous). The rise in the price of labor causes substitution of capital for labor. But the technical change bias increases the marginal productivity of labor relative to capital. In this case, if the technical change is sufficiently non-neutral, the "true" measure of technical change could fall outside of the index number "brackets"" (Ruttan 2001: 57–8).

5 New technologies and structural change

1   The reference to the behavioral theory of the firm, laid down by March and Simon (1958) and Cyert and March (1963), here is clear.

2   See the results of the analysis on pages 18–24.

3   Hicks (1976) provides a clear definition of the inducement hypothesis: "An induced invention is a change in technique that is made as a consequence of a change in prices (or, in general, scarcities); if the change in prices had not occurred, the change in technique would not have been made. I now like to think of a major technical change (one that we may agree to regard as autonomous, since, for anything that we are

concerned with, it comes from outside) as setting up what I call an Impulse. If the autonomous change is an invention which widens the range of technical possibilities, it must begin by raising profitability and inducing expansion; but the expansion encounters scarcities, which act as a brake. Some of the scarcities may be just temporary bottle-necks which in time can be removed; some, however, may be irremovable. Yet it is possible to adjust to either kind of scarcity by further changes in technical methods; it is these that are the true *induced inventions*. The whole story, when it is looked at in this way, is in *time*, and can be in history . . ." (Hicks, 1976/1982: 295 and 296) (italics as in the original text).

4 Irreversibility and switching costs are lower than in the case of general technologies, but not negligible in absolute terms. Irreversibility plays a key role in the general model of adjustment by means of technological change, as opposed to standard textbook technical change along a given isoquant and in a given map of isoquant, with no innovation, which consists of a simple change in factor intensity (Antonelli 1995, 1999a and 2001a).

5 In other models of this kind only changes in fixed capital where assumed to yield switching costs. See Antonelli 2001a.

6 Appropriate tuning of the parameters of equation (1) can express a range of conditions including the case in which switching costs depend almost exclusively upon the required changes in fixed capital, or in human capital, or in both.

7 In this model the firm considers the possibility of introducing new technologies in all possible technical directions. The direction of the innovation activity is not bound by the techniques in place. Localized learning takes place in the technique, defined in terms of input intensity, in place at each point in time, but it makes it possible to move in all directions so as to reshape the map of isoquants.

8 The metrics of technological change is defined in terms of rates of total factor productivity, while the metrics of technical change is provided by equation (1).

9 Respectively when the case for output maximization or cost minimization applies.

10 Paul David long ago suggested that the de-coupling of the inducement to innovate from the inducement of the direction of technological change was a fertile area of investigation. Little work however has been made since then along these lines. See David: "As soon as one is ready to discard the neoclassical conception of technological progress which insists that innovation and factor substition be viewed as logically distinct phenomena, there is no longer any great difficulty in taking an important step toward this proximate objective. Specifically it becomes possible to indicate how the realized factor-saving bias of 'changes in the state of technical arts' may come under the influence of factor-prices-directly, as well as indirectly through the medium of choice of technique decisions. In regard to the latter, we may for the present purposes eschew less orthodox 'behavioral' approaches to the decision making of firms; the prevailing structure of input prices will therefore continue to be cast in the governing role assigned to them by the traditional theory of rational, cost-minimizing firm" (David 1975: 57–8; see preliminary attempts to elaborate this point in Antonelli 1989 and 1990).

6 Industrial dynamics and technological change

1 See in Chapter 2 both the analysis on pages 10–11 of the results of the application of the product life cycle and the results of the analysis of the industrial dynamics along the technological trajectory summarized on pages 18–24.

2 These delays in the adjustement of relative prices help explaining the so-called productivity paradox and provide a complementary explanantion to the so-called "David's delay" (See David 1990 and 2001; Jorgenson 2001).

2 See the paragraph 5.4

7 The dynamics of factors markets and technological change

1   This analysis can be replicated with respect to the supply of savings and financial capital in an economic system. The introduction of new capital intensive technologies is likely to increase not only the demand for specific capital goods, but also the demand for financial capital and hence savings with effects on real interest rates. Again it seems clear that the provision of capital in each economic system is influenced by the quality of financial markets and in the international arena by the monetary and currency policies of the main financial institutions. A country able to increase local savings by means of budgetary policy and to attract international savings by means of the quality of local financial insitutions and an international currency policy which favor expectations about the re-evaluation of the local currency can keep real interest rates at low levels and take better advantage of the scope for total factor productivity levels associated with the introduction of new capital intensive technologies. An interesting trade-off emerges here between the positive effects of the high levels of each currency with respect to the levels of real interest rates and the negative effects of high exchange rates on exports and hence international competitiveness.

2   Complementary results are likely to be obtained by reducing the working time only if the supply of skilled labor is increased by *reskilling* policies.

3   Economic history, however, provides contrasting evidence on the matter. Recent analyses of the effects of Italian economic integration, at the time of the unification of the Italian Kingdom, suggest that the economy of the Kingdom of Naples was destroyed by the exposure to competition of the northern regions (Zamagni, 1990). Such (prolonged) failure has been mainly interpreted in terms of inefficiency and lack of flexibility of local factor markets. Yet the economic system of the Southern Kingdom in the pre-unification period was characterized by fast rates of growth. An effective industrial base was already in place with large firms specializing in highly labor intensive activities. The evidence recalls that the largest mill in the Italian regions was in fact near Naples and had over 10,000 employees. The collapse of the southern economy in the post-unification period was in fact too fast and radical to support the hypothesis of an economic system characterized by the lack of entrepreneurial energies and a generalized rent-seeking behaviour. This evidence might be interpreted as the result of the integration of a labor-abundant economy with a capital-abundant one, each of which using respctively a labor-intensive and capital-intensive technology.

4   In the Italian case, Piedmont and Lombardy.

5   The implications of this analysis with respect to past experiences of economic integration such as the unification of the Italian kingdom are far-reaching. The real weakness of the southern regions now seems to consist in the delay in adjusting the local technology to the changes in the conditions of the factor markets. While the static efficiency declined because of contingent changes in the factors markets, the southern regions lacked appropriate levels of dynamic efficiency.

8 Product innovation and barriers to entry

1   Let us assume an ordinary Cobb-Douglas utility function:

(1) $U = (Q_x)^a (Q_y)^b$

constrained maximization, with a given budget (R) and relative prices P_x and P_y, leads to a demand equation of the type:

(2) $(P_x) = R / P_x (1 + b/a)$

In the demand equation (2) the effects of the relative preference for the good x with respect to the good y is well shown by the ratio b/a. The effects of the prices for the substitute, however, is not directly reflected.

2   Valid for the domain $P_y < R$

3   The assumption that γ is positive but also that $\gamma < 1$ seems plausible considering that substitution is less and less effective with the reduction of the prices for substitutes and hence the levels of the revenue effects.

10 Feedbacks, path-dependence and evolution

1   The cliometric approach was born and has grown up quite successfully in the attempt to extend the domain of economics and apply to the analysis of economic history the basic notions of economics such as constrained optimization and most of all systemic interdependence. In so doing, however, cliometrics has made it possible to better identify and grasp the limitations and shortcomings of standard economics in dealing with historic time and dynamics at large. The reverse process is now at work, one where the basic categories of historic analysis, such as irreversibility and sequence, are applied to the analysis of economic processes.

12 Policy implications

1   It seems very important and necessary that there should be a systematic analysis of the characters and the effects of the regional distribution of the funds made available by the European Commission for research activities and their effects in terms of both concentration and local (as opposed to continental) effects in terms of the rate and direction of technological change.

13 Conclusions

1   In so doing this book elaborates upon the metaphor of "elastic barriers" suggested by Paul David (1975).

2   When the notions of irreversibility, local externalities, sequential change, and of innovation, viewed as a creative reaction, hence of path-dependence, are put in place and integrated into basic economics, a fully articulated post-Walrasian approach can be elaborated, one where the central role of the markets as mechanisms for the creation and distribution of incentives is recognized and emphsized. In such an approach, however, the welfare attributes of equilibrium are no longer valid. Equilibrium itself is questioned. Sequences of possible equilibria can be identified and traced. This approach provides a context in which economic analysis paves the way to economic policy.

Bibliography

Abbate, J. (1999) *Inventing the Internet*, Cambridge, MA, MIT Press.

Abernathy, W.J. and Utterback, J.M. (1978) "Patterns of innovation in technology", *Technology Review* 80, 40–47.

Abramovitz, M. (1956) "Resources and output trends in the US since 1870", *American Economic Review* 46, 5–23.

Abramovitz, M. (1979) "Rapid growth potential and its realization: The experience of the capitalist economies in the postwar period", in Malinvaud, E. (ed.), *Economic growth and resources*, London, Macmillan.

Abramovitz, M. (1989) *Thinking about growth*, Cambridge, Cambridge University Press.

Acemoglu, D. (1998) "Why do new technologies complement skills? Directed technical change and wage inequality", *Quarterly Journal of Economics* 113, 1055–1089.

Acemoglu, D. and Zilibotti, F. (2001) "Productivity differences", *Quarterly Journal of Economics* 116, 563–606.

Acs, Z.J. and Audretsch, D.B. (1990) *Innovation and small firms*, Cambridge, MA, MIT Press.

Agarwala, A.N. and Singh, S.P. (eds) (1958) *The economics of underdevelopment*, Oxford, Oxford University Press.

Aghion, P. and Howitt, P. (1998) *Endogenous growth theory*, Cambridge, MA, MIT Press.

Ahmad, S. (1966) On the theory of induced invention, *Economic Journal* 76, 344–357.

Allen, T.J. and Scott Morton, M.S. (eds) (1994) *Information technology and the corporation of the 90's*, Oxford, Oxford University Press.

Amendola, M. (1976) *Macchine produttività e progresso tecnico*, Milan, ISEDI.

Antonelli, C (ed.) (1988) *New information technology and industrial dynamics*, Boston and Dordrecht, Kluwer Academic.

Antonelli, C. (1989) "A failure inducement model of research and development expenditure: Italian evidence from the early 1980's", *Journal of Economic Behavior and Organization* 12, 159–180.

Antonelli, C. (1990) "Induced adoption and externalities in the regional diffusion of new information technology", *Regional Studies* 24, 31–40.

Antonelli, C. (1991) *The diffusion of advanced telecommunications in developing countries*, Paris, OECD.

Antonelli, C. (ed.) (1992) *The economics of information networks*, Amsterdam, Elsevier.

Antonelli, C. (1993a) "Externalities and complementarities in telecommunications dynamics", *International Journal of Industrial Organization* 11, 437–448.

Antonelli, C. (1993b) "The dynamics of technological interrelatedness. The case of information and communication technologies", in Foray, D. and Freeman, C. (eds), *Technology and the Wealth of Nations*, London, Pinter.

Antonelli, C. (1995) *The economics of localized technological change and industrial dynamics*, Boston, Kluwer.

Antonelli, C. (1999a) *The microdynamics of technological change*, London, Routledge.

Antonelli, C. (ed.) (1999b) *Conoscenza tecnologica*, Torino, Edizioni della Fondazione Giovanni Agnelli.

Antonelli, C. (2001a) *The microeconomics of technological systems*, Oxford, Oxford University Press.

Antonelli, C. (2001b) "Cambiamento tecnologico: Prezzi relativi e dinamiche della crescita produttiva", in *Tecnologia e società*, Accademia dei Lincei, Roma.

Antonelli, C. (2002) "The governance of knowledge commons", in Geuna, A., Salter, A. and Steinmueller, W.E. (eds) *Science and innovation. Rethinking the rationales for funding and governance*, Cheltenham, Edward Elgar.

Antonelli, C. (2002a) "Innovation and structural change", *Economie Appliqueé* 55.

Antonelli, C. (2003) "The digital divide: Understanding the economics of new information and communication technology in the global economy", *Information Economics and Policy* 15.

Antonelli, C. and Calderini, M. (2001) *Le misure della ricerca. Attività scientifica a Torino*, Torino, Edizioni della Fondazione Giovanni Agnelli.

Antonelli, C. and Quèrè, M. (2002) "The governance of interactive learning within innovation systems", *Urban Studies* 39, 1051–1063.

Antonelli, C., Gaffard, J.L. and M. Quéré (2003) "Interactive learning and technological knowledge: The localized character of innovation processes", in Rizzello, S. (ed.), *Cognitive paradigms in economics*, Routledge, London.

Antonelli, C., Petit, P. and Tahar, G. (1992) *The economics of industrial modernization*, Cambridge, Academic Press.

Argyres, N.S. (1995) "Technology strategy governance structure and interdivisional coordination", *Journal of Economic Behavior and Organization* 28, 337–358.

Argyres, N.S. and Liebeskind, J.P. (1998) "Privatizing the intellectual commons: Universities and the commercialization of biotechnology", *Journal of Economic Behavior and Organization* 35, 427–454.

Arora, A., Fosfuri, A. and Gambardella, A. (2001) *Markets for technology*, Cambridge, MA, MIT Press.

Arrow, K.J. (1962a) "Economic welfare and the allocation of resources for invention", in Nelson, R.R. (ed.) *The rate and direction of inventive activity: Economic and social factors*, Princeton, Princeton University Press for N.B.E.R.

Arrow, K.J. (1962b) "The economic implications of learning by doing", *Review of Economic Studies* 29, 155–173.

Arrow, K.J. (1969) "Classificatory notes on the production and transmission of technical knowledge", *American Economic Review* (P&P) 59, 29–35.

Arrow, K.J. (2000) "Increasing returns: Historiographic issues and path dependence", *European Journal of History of Economic Thought* 7, 171–180.

Arthur, B. (1989) "Competing technologies increasing returns and lock-in by small historical events", *Economic Journal* 99, 116–131.

Arthur, B. (1994) *Increasing returns and path dependence in the economy*, Ann Arbor, Michigan University Press.

Asimakopulos, A. and Weldon, J.C. (1963) "The classification of technical progress in models of economic growth", *Economica* 30, 372–386.

Atkinson, A.B. and Stiglitz, J.E. (1969) "A new view of technological change", *Economic Journal* 79: 573–578.

Attewell, P. (1994) "Information technology and the productivity paradox", in Harris, D. (ed.), *Organizational linkages: Understanding the productivity paradox*, Washington, National Academy of Sciences Press.

Audretsch, D.B. (1995) *Innovation and industry evolution*, Cambridge, MIT Press.

Audretsch, D.B. and Feldman, M. (1996) "Spillovers and the geography of innovation and production", *American Economic Review* 86, 630–640.

Audretsch, D.B. and Klepper. S. (eds) (2000) *Innovation evolution of industry and economic growth*, Cheltenham, Edward Elgar.

Audretsch, D.B. and Stephan, P.E. (1996) "Company-scientist locational links: The case of biotechnology", *American Economic Review* 86, 641–652.

Audretsch, D.B., Menkveld, A.J. and Thurik, A.R. (1996) "The decision between internal and external R&D", *Journal of Institutional and Theoretical Economics* 152, 519–530.

Autant-Bernard, C. (2001) "Science and knowledge flows: Evidence from the French case", *Research Policy* 30, 1069–1078.

Autor, D.H., Katz, L.F. and Krueger, A.B. (1998) "Computing inequality: Have computers changed the labor market?", *Quarterly Journal of Economics* 113, 1169–1213.

Bartel, A.P. and Lichtenberg, F.R. (1987) "The comparative advantage of educated workers in implementing new technology", *Review of Economics and Statistics* 69, 1–11.

Bartel, A.P. and Lichtenberg, F.R. (1990) "The impact of age of technology on employee wages", *Economics of Innovation and New Technology* 1, 1–17.

Baumol, W.J. (1982) "Contestable markets: An uprising in the theory of industry structure", *American Economic Review* 72, 1–15.

Baumol, W.J. (1986) "Productivity growth convergence and welfare: What the long-run data show", *American Economic Review* 78, 1072–1085.

Baumol, W.J., Nelson, R.R. and Wolff, E.N. (eds) (1994) *Convergence of productivity. Cross-national studies and historical evidence*, Oxford, Oxford University Press.

Berman, E., Bound, J. and Machin, S. (1998) "Implication of skilled-biased technological change: International evidence", *Quarterly Journal of Economics*, 113, 1245–1279.

Besomi, D. (1999) "Harrod on the classification of technological progress. The origin of a wild-goose chase", *BNL Quarterly Review* 208, 95–118.

Bijker, W.E., Hughes, T.P. and Pinch, T. (eds) (1987) *The social construction of technological systems*, Cambridge, MA, MIT Press.

Binswanger, H.P. and Ruttan, V.W. (eds), (1978) *Induced innovation: Technology institutions and development*, Baltimore, Johns Hopkins University Press.

Blaug, M. (1963) "A survey of the theory of process-innovations", *Economica* 30, 13–32.

Bonte, W. (2003) "R&D and productivity: Internal vs. external R&D. Evidence from West German manufacturing industries", *Economics of Innovation and New Technology* 12, forthcoming.

Bresnahan, T.F. (1999) "Computerization and wage dispersion: An analytical interpretation", *Economic Journal* 109, F390–F415.

Bresnahan, T.F., Brynjolfsson, E. and Hitt, L. (1999) "Information technology and recent changes in work organization increase the demand for skilled labor", in Blair, M. and Kochan, T. (eds) *The new relationship: Human capital in the American corporation*, Washington, The Brookings Institution.

Bresnahan, T.F., and Traitenberg, M. (1995) "General purpose technologies: "Engines of growth"?" *Journal of Econometrics* 65, 83–108.

Broadberry, S. (1997) *The productivity race. British manufacturing in international perspective, 1850–1990*, Cambridge, Cambridge University Press.

Brown, M. (1966) *On the theory and measurement of technological change*, Cambridge, Cambridge University Press.

Brynjolfsson, E. and Hitt, L. (1995) "Information technology as a factor of production: The role of difference among firms", *Economics of Innovation and New Technology* 3, 183–200.

Brynjolfsson, E. and Kahin, B. (eds) (2001) *Understanding digital economy*, Cambridge, MA, MIT Press.

Brynjolfsson, E. and Yang, S. (1996) "Information technology and productivity: A review of literature", *Advances in Computers* 43, 179–214.

Callon, M. (1989) *La science et ses reseaux. Genèse et circulation des faits scientifiques*, Paris, La Decouverte.

Carlsson, B. (ed.) (1995) *Technological systems and economic performance: The case of factory automation*, Boston, Kluwer Academic Publishers.

Carlsson, B. (1998) "On and off the beaten path: The evolution of four technological systems in Sweden", *International Journal of Industrial Organization* 15, 775–800.

Carlsson, B. (ed.) (2002) *Technological systems in the bio industries. An international study*, Boston, Kluwer Academic Publishers.

Carlsson, B. and Eliasson, G. (1994) "The nature and importance of economic competence", *Industrial and Corporate Change* 3, 687–712.

Carlsson, B. and Stankiewitz, R. (1991) "On the nature, function and composition of technological systems", *Journal of Evolutionary Economics* 1, 93–118.

Carroll, G.R. and Teece, D.J. (eds) (1999) *Firms markets and hierarchies. The transaction cost economics perspective*, Oxford, Oxford University Press.

Caselli, F. (1999) "Technological revolutions", *American Economic Review* 89, 78–102.

Caselli, F. and Coleman, W.J. (2001) "The US structural transformation and regional convergence: A reinterpretation", *Journal of Political Economy* 109, 584–616.

Caves, R. (2000) *Creative industries. Contracts between art and commerce*, Cambridge, MA, Harvard University Press.

Chalmers, R.G. (1988) *Applied production analysis: A dual approach*, Cambridge, Cambridge University Press.

Chari, V.V. and Hopenhayn, H. (1991) "Vintage human capital growth and the diffusion of new technology", *Journal of Political Economy* 99, 1142–1165.

Chenery, H. and Syrquin, M. (with the assistance of Elkinton, H.) (1975) *Patterns of development*, Oxford, Oxford University Press for the World Bank.

Chenery, H., Robinson, S. and Syrquin, M. (1986) *Industrialization and growth*, Oxford, Oxford University Press for the World Bank.

Cohendet, P., Llerena, P., Stahn, H. and Umbhauer, G. (1998) *The economics of networks*, Berlin, Springer-Verlag.

Cooper, D.P. (2001) "Innovation and reciprocal externalities: Information transmission via job mobility", *Journal of Economic Behavior and Organization* 45, 403–425.

Cowan, R., David, P.A. and Foray, D. (2000) "The explicit economics of knowledge codification and tacitness", *Industrial and Corporate Change* 9, 211–253.

Cozzi, T. and Marchionatti, R. (eds) (2000) *Piero Sraffa's political economy. A centenary estimate*, London, Routledge.

Crandall, R.W. and Flamm, K. (eds) (1989) *Changing the rules: Technological change, international competition and regulation in communications*, Washington, The Brookings Institution.

Cusumano, M.A. and Selby, R.H. (1995) *Microsoft secrets*, New York, The Free Press.

Cyert, R.M. and March, J.C. (1963) *A behavioral theory of the firm*, Englewood Cliffs, Prentice-Hall.

Dasgupta, P. and Stiglitz, J.E. (1980) "Industrial structure and the nature of innovative activity", *Economic Journal* 90, 266–293.

David, P.A. (1975) *Technical choice innovation and economic growth*, Cambridge, Cambridge University Press,

David, P.A. (1985) "Clio and the economics of QWERTY", *American Economic Review* 75, 332–337.

David, P.A. (1987) "Some new standards for the economics of standardization in the information age", in Dasgupta, P. and Stoneman, P. (eds), *Economic policy and technological performance*, Cambridge, Cambridge University Press.

David, P.A. (1990) "The dynamo and the computer: A historical perspective on the productivity paradox", *American Economic Review* (P&P) 80, 355–361.

David, P.A. (1992) "Heroes, herds and hysteresis in technological history", *Industrial and Corporate Change* 1, 129–179.

David, P.A. (1993) "Knowledge property and the system dynamics of technological change", *Proceedings of the World Bank Annual Conference on Development Economics*. Washington, The World Bank.

David, P.A. (1994) "Positive feedbacks and research productivity in science: Reopening another black box", in Granstrand, O. (ed.), *Economics and technology*, Amsterdam, Elsevier.

David, P.A. (1997a) "From market magic to calypso science policy. A review of Terence Kealey's 'The economic laws of scientific research'", *Research Policy* 26, 229–255.

David, P.A. (1997b) *Path dependence and the quest for historical economics: One more chorus of the ballad of QWERTY*, University of Oxford Discussion Papers in Economic and Social History, Number 20.

David, P.A. (1998) "Communication norms and the collective cognitive performance of 'Invisible Colleges'", in Barba Navaretti, G. *et al.* (eds), *Creation and the transfer of knowledge: Institutions and incentives*, Berlin, Heidelberg, New York, Springer-Verlag.

David, P.A. (2001) "Understanding digital technology's evolution and the path of measured productivity growth: Present and future in the mirror of the past", in Brynjolfsson, E. and Kahin, B. (eds).

David, P.A. and Foray, D. (1994) "The economics of EDI standards diffusion", in Pogorel, G. (ed.), *General telecommunications strategies and technological changes*, Amsterdam, Elsevier.

DeBresson, C. and Townsend, J. (1978) "Notes on the inter-industrial flow of technology in post-war Britain", *Research Policy* 7, 48–60.

DeLong, J.B. (1988) "Productivity growth convergence and welfare: Comment". *American Economic Review* 78, 1138–1154.

Domar, E.D. (1961) "On the measurement of technical change", *Economic Journal* 71, 739–755.

Dorfman, N. (1987) *Innovation and market structure: Lessons from the computer and semiconductor industry*, Cambridge, MA, Ballinger.

Dosi, G. (1982) "Technological paradigms and technological trajectories: A suggested interpretation of the determinants and directions of technological change", *Research Policy* 11, 147–162.

Dosi, G. (1988) "Sources procedures and microeconomic effects of innovation", *Journal of Economic Literature* 26, 1120–1171.

Dosi, G., Freeman, C., Nelson, R., Silverberg, G., Soete, L. (eds) (1988) *Technical change and economic theory*, London, Pinter.

Dumont, B. and Holmes, P. (2002) "The scope of intellectual property rights and their interface with competition law and policy: Divergent paths to the same goal", *Economics of Innovation and New Technology* 11, 149–162.

Duysters, G. (1996) *The dynamics of technical innovation. The evolution and development of information technology*, Cheltenham, Edward Elgar.

Eckaus, R.S. (1955) "The factor proportions problem in underveloped areas", *American Economic Review*, 1955, in Agarwala, A.N. and Singh, S.P. (eds) (1958) *The economics of underdevelopment*, Oxford, Oxford University Press.

Edquist, C. (ed.) (1997) *Systems of innovation: Technologies institutions and organizations*, London, Pinter.

Elster, J. (1983) *Explaining technical change*, Cambridge, Cambridge University Press.

Eltis, W.A. (1971) "The determination of the rate of technical progress", *Economic Journal* 61, 502–524.

Etzkowitz, H. and Leydesdorff, L. (2000) "The dynamics of innovation: From national systems and 'Mode 2' to triple helix of university-industry-government relations", *Research Policy* 29, 109–123.

Evenson, R.E. and Westphal, L.E. (1995) "Technological change and technology strategy", in Behman, J. and Srinivasan, T.N. (eds), *Handbook of development economics*, Amsterdam, Elsevier.

Falk, M. and Seim, K. (2001) "The impact of information technology on high-skilled labor in services: Evidence from firm-level panel data", *Economics of Innovation and New Technology* 10, 289–324.

Fellner, W. (1961) "Two propositions in the theory of induced innovation", *Economic Journal* 71, 305–308.

Foray D. (2000) *L'économie de la connaissance*, Paris, La Decouverte.

Foss, N.J. (1997) *Resources, firms and strategies. A reader in the resource-based perspective*, Oxford, Oxford University Press.

Foss, N.J. (1998) "The resource-based perspective: An assessment and diagnosis of problems", *Scandinavian Journal of Management* 15, 1–15.

Foss, N. and Mahnke, V. (eds) (2000) *Competence governance and entrepreneurship*, Oxford, Oxford University Press.

Fransman, M. (1995) *Japan's computer and communications industry*, Oxford, Oxford University Press.

Fransman, M. (1999) *Visions of innovation*, Oxford, Oxford University Press.

Freeman, C. (1991) "Networks of innovators: A synthesis of research issues", *Research Policy* 20, 499–514.

Freeman, C. (1994) "The economics of technical change", *Cambridge Journal of Economics* 18, 463–514.

Freeman, C. and Louca, F. (2001) *As time goes by: From industrial revolution to information revolution*, Oxford, Oxford University Press.

Freeman, C., Clark, J. and Soete, L. (1892) *Unemployment and technical innovation. A study of long waves and economic development*, London, Pinter.

Friedman, M. (1992) "Do old fallacies ever die?", *Journal of Economic Literature* 30, 2129–2132.

Furubotn, E.G. (2001) "The new institutional economics and the theory of the firm", *Journal of Economic Behavior and Organization* 45, 133–153.

Garrouste, P. and Ioannides, S. (eds) (2001) *Evolution and path dependence in economic ideas. Past and present*, Cheltenham, Edward Elgar.

Gaskins, D.W. jr., (1971) "Dynamic limit pricing: Optimal limit pricing under threat of entry", *Journal of Economic Theory* 3, 306–322.

Geroski, P. (1991) *Market dynamics and entry*, Oxford, Basil Blackwell.

Geuna, A. (1999) *The economics of knowledge production*, Cheltenham, Edward Elgar.

Geuna, A., Salter, A. and Steinmueller, W.E. (eds) (2002) *Science and innovation. Rethinking the rationales for funding and governance*, Cheltenham, Edward Elgar.

Gibbons, M., Limoges, C., Nowotny, H., Schwarzman, S., Scott, P. and Trow, M. (1994) *The new production of knowledge: The dynamics of research in contemporary societies*, London, SAGE Publications.

Goddard, F.O. (1970) "Harrod-neutral economic growth with Hicks-biased technological progress", *Southern Economic Journal* 26, 300–308.

Goldin, C. and Katz, L. (1998) "The origins of the technology-skill complementarity", *Quarterly Journal of Economics* 113, 693–732.

Goldin, C., and Katz, L.F. (1996) "Technology, skill, and the wage structure", *American Economic Journal* 86, .252–57

Gould, S.J. (2002) *The structure of evolutionary theory*, Cambridge, MA, Harvard University Press.

Greenan, M. and Mairesse, J. (2000) "Computers and productivity in France: Some evidence", *Economics of Innovation and New Technology* 9, 275–315.

Griliches, Z. (1956) "Hybrid corn: An exploration in the economics of technical change", *Econometrica* 25, 501–522.

Griliches, Z. (1969) "Capital-skill complementarity", *Review of Economics and Statistics* 51, 465–468.

Griliches, Z. (1979) "Issues in assessing the contribution of research and development to productivity growth", *Bell Journal of Economics* 10, 92–116.

Griliches, Z. (1992) "The search for R&D spillovers", *Scandinavian Journal of Economics* 94, 29–47.

Griliches, Z. (1997) "The discovery of the residual", *Journal of Economic Literature* 34, 1324–1330.

Grindley, P. (1995) *Standards strategy and policy: Cases and stories*, Oxford, Oxford University Press.

Grossman, S.J. and Hart, O.D. (1986) "The costs and benefits of ownership: A theory of vertical integration", *Journal of Political Economy* 94, 691–719.

Habakkuk, H.J. (1962) *American and British technology in the nineteenth century*, Cambridge, Cambridge University Press.

Harrod, R.F. (1939) "An essay in dynamic theory", *Economic Journal* 49, 14–33.

Hart, O.D. (1995) *Firms contracts and financial structure*, Clarendon Lectures in Economics, Oxford, Oxford University Press.

Hayek, F.A. (1945) "The use of knowledge in society", *American Economic Review* 35, 519–530.

Heertje, A. (1973) *Economics and technological change*, London, Weidenfeld and Nicolson.

Helpman, E. (ed.) (1998) *General purpose technologies and economic growth*, Cambridge, MA, MIT Press.

Henderson, J.M. and Quandt, R.E. (1971 and 1980) *Microeconomic theory. A mathematical approach*, New York, McGraw-Hill Book Company.

Hicks, J.R. (1932) *The theory of wages*, London, Macmillan.

Hicks, J.R. (1976) "Time in Economics", in Tang, A.M., *Evolution welfare and time in economics*. Quotes are drawn from Hicks, J.R. *Collected essays on economic theory*, Volume II, Oxford, Basil Blackwell.

Holmström, B. and Roberts, J. (1998) "The boundaries of the firm revisited", *Journal of Economic Perspectives* 12 , 73–94.

Howells, J. (1999) "Research and technological outsourcing", *Technology Analysis & Strategic Management* 11, 17–29.

Jorgenson, D.W. (2001) "Information technology and the US economy", *American Economic Review* 91, 1–33.

Jorgenson, D.W. and Stiroh, K.J. (1995) "Computers and growth", *Economics of Innovation and New Technology* 3, 295–316.

Jorgenson, D.W. and Stiroh, K.J. (2000) "Raising the speed limit: US economic growth in the information age", *Brookings Papers on Economic Activity* 1, 125–211.

Katz, J. (ed.) (1987) *Technology generation in Latin American manufacturing industries*, London, St Martin's Press.

Katz, M. and Shapiro, C. (1985) "Network externalities, competition and compatibility", *American Economic Review* 75, 424–440.

Kennedy, C. (1966) "Induced bias and the theory of distribution", *Economic Journal* 76, 541–547.

Kennedy, C. and Thirwall, A.P. (1972) "Survey in applied economics: Technical progress", *Economic Journal* 82, 11–72.

Kiley, M.T. (1999) "The supply of skilled labor and skilled biased technological progress", *Economic Journal* 109, 708–724.

Kingston, W. (2001) "Innovation needs patents reform", *Research Policy* 30, 403–423.

Klepper, S. (1996) "Entry, exit and innovation over the product life cycle", *American Economic Review* 86, 562–583.

Klepper, S. and Graddy, E. (1990) "The evolution of new industries and the determinants of market structure", *Rand Journal of Economics* 21, 27–44.

Klepper, S. and Miller, J.H. (1995) "Entry, exit and shakeouts in the United States in new manufactured products", *International Journal of Industrial Organization* 13, 567–591.

Klepper, S. and Simons, K.L. (20000) "The making of an oligopoly: Firm survival and technological change in the evolution of the US tire industry", *Journal of Political Economy* 108, 728–760.

Kline, S.J. and Rosenberg, N. (1986) "An overview of innovation", in Landau, R. and Rosenberg, N. (eds), *The positive sum strategy*, Washington, National Academy Press.

Krueger, A. (1993) "How computers have changed the wage structure: Evidence from microdata, 1984–89", *Quarterly Journal of Economics* 108, 33–59.

Krugman, P.R. (1991) *Geography and trade*, Cambridge, MA, MIT Press.

Krugman, P.R. (1995) *Development geography and economic theory*, Cambridge, MA, MIT Press.

Kuhn, T. (1962) *The strucure of scientific revolutions*, Chicago, University of Chicago Press.

Lall, S. (1987) *Learning to industrialize: The acquisition of technological capability by India*, London, Macmillan.

Lamberton, D. (ed.) (1971) *Economics of information and knowledge*, Harmondsworth, Penguin.

Landau, R., Taylor, T. and Wright, G. (1996) *The mosaique of economic growth*, Stanford, Stanford University Press.

Langlois, R.N. (ed.) *Economics as a process: Essays in the new institutional economics*, Cambridge, Cambridge University Press.

Latour, B. (1987) *Science in action. How to follow scientists and engineers in society*, Cambridge, MA, Harvard University Press.

Lichtenberg, F. (1995) "The output contributions of computer equipment and personnel: A firm level analysis", *Economics of Innovation and New Technology* 3, 201–217.

Loasby, B.J. (1999) *Knowledge institutions and evolution in economics*, London, Routledge.

Loyd-Ellis, H. (1999) "Endogenous technological change and wage inequality", *American Economic Review* 89, 47–77.

Lundvall, B.A. (1985) *Product innovation and user-producer interaction*, Aalborg, Aalborg University Press.

Lypsey, R. (2002) "Some implications of endogenous technological change for technology policies in developing countries", *Economics of Innovation and New Technology* 11, 321–352.

Lypsey, R., Bekar, C. and Carlaw, K. (1998) "General purpose technologies: What requires explanation", in Helpman, E. (ed.) *General purpose technologies and economic growth*, Cambridge, MA, MIT Press.

Machin, S. and van Reenen, J. (1998) "Technology and changes in skill structure: Evidence from an international panel of industries", *Quarterly Journal of Economics* 113, 1215–1244.

Machlup, F. (1962) *The production and distribution of knowledge in the United States*, Princeton, Princeton University Press.

Madden, G. and Savage, S.J. (2001) "Productivity growth and market structure in telecommunications", *Economics of Innovation and New Technology* 10, 493–512.

Malerba, F. (1985) *The semiconductor industry. The economics of rapid growth and decline*, London, Pinter.

Malerba, F. (1992) "Learning by firms and incremental technical change", *Economic Journal* 102, 845–859.

Malerba, F. (1996) "Schumpeterian patterns of innovation", *Cambridge Journal of Economics* 19, 47–65.

Mansell, R. (1994) *The new telecommunications*, New York, Sage.

Mansell, R. (2001) "Digital opportunities and the missing link for developing countries", *Oxford Journal of Economic Policy* 17, 282–295.

Mansell, R. and Wehen, U. (eds) (1998) *Knowledge societies: Information technology for sustainable development*, Oxford, Oxford University Press for the United Nations Commission on Science and Technology for Development.

March, J.C. and Simon, H.A. (1958) *Organizations*, New York, John Wiley & Sons.

Marchesi, M. (2002) "Essays in the Economics of the Information Industry", PhD Thesis, School of Economic Studies, University of Manchester.

Marchionatti, R. (1999) "On Keynes' animal spirits", *Kyklos* 52, 415–439.

Marchionatti, R. (2000) "Sraffa and the criticism of Marshall in the 1920s", in Cozzi, T. and Marchionatti, R. (eds), *Piero Sraffa's political economy. A centenary estimate*, Routledge, London.

Marmolo, E. (1999) "A constitutional theory of public goods", *Journal of Economic Behavior and Organization* 38, 27–42.

Martin, S. (1993) *Advanced industrial economics*, Oxford, Basil Blackwell.

Mazzoleni, R. and Nelson, R.R. (1998) "The benefits and the costs of strong patent protection: A contribution to the current debate", *Research Policy* 27, 273–284.

Menard, C. (ed.) (2000) *Institutions contracts and organizations. Perspectives from new institutional economics*, Aldershot, Edward Elgar.

Merton, R. (1973) *The sociology of science: Theoretical and empirical investigations*, Chicago, University of Chicago Press.

Metcalfe, J.S. (1981) "Impulse and diffusion in the study of technical change", *Futures* 13, 347–359.

Metcalfe, J.S. (1995) "Technology systems and technology policy in historical perspective", *Cambridge Journal of Economics* 19, 25–47.

Metcalfe, S. (1995) "The economic foundation of technology policy: Equilibrium and evolutionary perspectives", in Stoneman, P. (ed.) *Handbook of the economics of innovation and technological change*, Oxford, Basil Blackwell.

Metcalfe, J.S. (1997) *Evolutionary economics and creative destruction*, London, Routledge.

Misa, T.J. (1995) "Retrieving sociotechnical change from technological determinism", in Smith, M. R. and Marx, L. (eds), *Does technology drive history ? The dilemma of technological determinism*, Cambridge, MA, MIT Press.

Modigliani, F. (1958) "New developments on the oligopoly front", *Journal of Political Economy* 46, 215–232.

Mokyr, J. (1990) *The lever of riches. Technological creativity and economic progress*, Oxford, Oxford University Press.

Momigliano, F. (1975) *Economia industriale e teoria dell'impresa*, Bologna, Il Mulino.

Mowery, D.C. (1996) *The international computer software industry: A comparative study of industry evolution and structure*, Oxford, Oxford University Press.

Nadiri, M.I. and Nandi, B. (2001) "Benefits of communications infrastructure capital in US economy", *Economics of Innovation and New Technology* 10, 89–108.

Nelson, R.R. (1968) "A diffusion model of international productivity differences in manufacturing industry", *American Economic Review* 58, 1219–1248.

Nelson, R.R. (ed.) (1993) *National systems of innovation*, Oxford, Oxford University Press.

Nelson, R.R. and Sampat, B. (2001) "Making sense of institutions as a factor shaping economic performance", *Journal of Economic Behavior and Organization* 44, 31–54.

Nelson, R.R. and Winter, S.G. (1982) *An evolutionary theory of economic change*, Cambridge, MA, Harvard University Press.

Nooteboom, B. (2000) *Learning and innovation in organizations and economics*, Oxford, Oxford University Press.

Pasinetti, L.L. (1962) "Rate of profit and income distribution in relation to the rate of growth", *Review of Economic Studies* 29, 267–279.

Pasinetti, L.L. (1981) *Structural change and economic growth*, Cambridge, Cambridge University Press.

Pavitt, K. (1984) "Sectoral patterns of technical change: Towards a taxonomy and a theory", *Research Policy* 13, 343–375.

Pavitt, K. (2000) *Technology, management and systems of innovation*, Cheltenham, Edward Elgar.

Pavitt, K., Robson, M. and Townsend, J. (1989) "Technological accumulation, diversification and organisation in UK companies, 1945–83", *Management Science* 35, 81–99.

Penrose, E. (1959) *The theory of the growth of the firm*, Oxford, Basil Blackwell.

Pianta, M. and Vivarelli, M. (1999) "Employment dynamics and structural change in Europe", in Fagerberg, J., Guerrieri, P. and Verspagen, B. (eds), *The economic challenge for Europe: Adapting to innovation-based growth*, Cheltenham, Edward Elgar, 83–105.

Polanyi, M. (1958) *Personal knowledge. Towards a post-critical philosophy*, London, Routledge & Kegan Paul.

Polanyi, M. (1966) *The tacit dimension*, London, Routledge & Kegan Paul.

Quah, D. (2001) "The weightless economy in economic development", in Pohjola, M. (ed.), *Information technology productivity, and economic growth: International evidence and implications for economic development*, Oxford, Oxford University Press.

Robinson, J. (1937) "The classification of inventions", *Review of Economic Studies* 6, 139–142.

Romer, P.M. (1986) "Increasing returns and long run growth", *Journal of Political Economy* 94, 1002–1037.

Romer, P.M. (1994) "Endogenous technological change", *Journal of Political Economy* 98, 71–103.

Rosenberg, N. (ed.) (1971) *The economics of technological change*, Harmondsworth, Penguin.

Rosenberg, N. (1976) *Perspectives on technology*, Cambridge, Cambridge University Press.

Rosenberg, N. (1982) *Inside the black box. Technology and economics*, Cambridge, Cambridge University Press.

Rosenberg, N. (2000) *Schumpeter and the endogeneity of technology. Some American perspectives*, London, Routledge.

Ruttan, V.W. (1997) "Induced innovation evolutionary theory and path dependence: Sources of technical change", *Economic Journal* 107, 1520–1529.

Ruttan, V.W. (2001) *Technology growth and development. An induced innovation perspective*, Oxford, Oxford University Press

Rycroft, R.W. and Kash, D.E. (1999) *The complexity challenge*, London, Pinter.

Salter, W.E.G. (1960) *Productivity and technical change*, Cambridge, Cambridge University Press.

Sato, R. and Suzawa, G.S. (1983) *Research and productivity. Endogenous technical change*, Boston, Auburn House Publishing Company.

Saul, S.B. (ed.) (1970) *Technological change: The United States and Britain in the 19th Century*, London, Methuen.

Scherer, F.M. (1984) *Innovation and growth: Schumpeterian perspectives*, Cambridge, MA, MIT Press.

Scherer, F.M. (1992) *International high-technology competition*, Cambridge, MA, Harvard University Press.

Scherer, F.M. (1999) *New perspectives on economic growth and technological innovation*, Washington, Brookings Institution Press.

Schmookler, J. (1966) *Invention and economic growth*, Cambridge, MA, Harvard University Press.

Schreyer, P. (2001) "Information and communication technology and the measurement of volume output and final demand", *Economics of Innovation and New Technology* 10, 339–376.

Schumpeter, J.A. (1928) "The instability of capitalism", *Economic Journal* 38, 361–386.

Schumpeter, J.A. (1936) *The theory of economic development*, Cambridge, MA, Harvard University Press.

Schumpeter, J.A. (1942) *Capitalism, socialism and democracy*, London, Unwin.

Shapiro, C. and Varian, H. (1999) *Information rules*, Boston, Harvard Business School Press.

Simon, H.A. (1947) *Administrative behavior*, New York, The Free Press.

Simon, H.A. (1962) "The architecture of complexity", *Proceedings of the American Philosophical Society* 106, 467–482. Quotes are referred to Simon (1969).

Simon, H.A. (1969) *The sciences of artificial*, Cambridge, MA, MIT Press.

Simon, H.A. (1982) *Metaphors of bounded rationality. Behavioral economics and business organization*, Cambridge, MA, MIT Press.

Smith, M.R. and Marx, L. (eds) (1995) *Does technology drive history? The dilemma of technological determinism*, Cambridge, MA, MIT Press.

Solow, R.M. (1956) "A contribution to the theory of economic growth", *Quarterly Journal of Economics* 71, 65–94.

Solow, R.M. (1957) "Technical change and the aggregate production function", *Review of Economics and Statistics* 39, 312–320.

Solow, R.M. (1987) "We'd better watch out", *New York Review of Books*, July 12.

Steedman, I. (ed.) (1988) *Sraffian economics*, Aldershot, Edward Elgar.

Stiglitz, J.E. (1994) "Economic growth revisited", *Industrial and Corporate Change* 3, 65–110.

Stiroh, K.J. (1998) "Computers productivity and input substitution", *Economic Inquiry* 36, 175–191.

Stoneman, P. (1976) *Technological diffusion and the computer revolution*, Cambridge, Cambridge University Press.

Stoneman, P. (1983) *The economic analysis of technical change*, Oxford, Oxford University Press.

Stoneman, P. (1987) *The economic analysis of technology policy*, Oxford, Clarendon Press.

Stoneman, P. (ed.) (1995) *Handbook of the economics of innovation and technological change*, Oxford, Basil Blackwell.

Swann, P. (ed.) (1994) *New technologies and the firm: Innovation and competition*, London, Routledge.

Swann, P., Prevezer, M. and Stout, D. (eds) (1998) *The dynamics of industrial clustering*, Oxford, Oxford University Press

Sylos Labini, P. (1956) *Oligopolio e progresso tecnico*, Milano, Giuffrè. Translated into English (1962), *Oligopoly and technical progress*, Cambridge, MA, Harvard University Press.

Sylos Labini, P. (1984) *The forces of economic growth and decline*, Cambridge, MA, MIT Press.

Teece, D.J. (1986) "Profiting from technological innovation: Implications for integration collaboration licensing and public policy", *Research Policy* 15, 285–305.

Teece, D.J. (2000) *Managing intellectual capital*, Oxford, Oxford University Press.

Teitel, S. (1987) "Towards an understanding of technical change in semi-industrialized countries", in Katz, J. (ed), *Technology generation in Latin American manufacturing industries*, London, St Martin's Press.

Thesmar, D. and Thoenig, M. (2000) "Creative destruction and firm organization choice", *Quarterly Journal of Economics* 115, 1201–1237.

Thirtle, C.G. and Ruttan, V.W. (1987) *The role of demand and supply in the generation and diffusion of technological change*, London, Harwood Academic Publishers.

Utterback, J.M. (1994) *Mastering the dynamics of innovation*, Boston, Harvard Business School Press.

Vernon, R. (1966) "International investment and international trade in the product cycle", *Quarterly Journal of Economics* 80, 190–207.

Veugelers, R. and Cassiman, B. (1999) "Make and buy in innovation strategies: Evidence from Belgian manufacturing firms", *Research Policy* 28, 63–80.

Vivarelli, M. (1995) *The economics of technology and employment: Theory and empirical evidence*, Cheltenham, Edward Elgar.

Vivarelli, M. and Pianta, M. (2000) (eds), *The employment impact of innovation: Evidence and policy*, London, Routledge,

Vivarelli, M., Evangelista, R. and Pianta, M. (1996) "Innovation and employment: Evidence from Italian manufacturing", *Research Policy* 25, 1013–1026.

Von Hippel, E. (1988) *The sources of innovation*, London, Oxford University Press.

Wasserman, N.H. (1985) *From invention to innovation: Long distance telephone transmission at the turn of the century*, Baltimore, Johns Hopkins University Press.

Westphal, L.E. (1990) "Industrial policy in an export-propelled economy: Lessons from South Korea's experience", *Journal of Economic Perspectives* 4, 41–59.

Williamson, O.J. (1985) *The economic institutions of capitalism*, New York, Free Press.

Williamson, O.(1996) *The mechanisms of governance*, New York, Oxford University Press.

Wright, G. (1997) "Toward an historical approach to technological change", *Economic Journal* 107, 1560–1566.

Wyckoff, A.W. (1995) "The impact of computer prices on international comparisons of labour productivity", *Economics of Innovation and New Technology* 3, 277–293.

Zamagni, V. (1990) *Dalla periferia al centro. La seconda rinascita economica dell'Italia: 1861–1990*, Bologna, Il Mulino.

Zeira, J. (1998) "Workers, machines and economic growth", *Quarterly Journal of Economics* 113, 1091–1117.

Index

Printed in the United Kingdom
by Lightning Source UK Ltd.
115794UKS00004B/268-270

9 780415 406437